GENERAL BIOLOGY
LABORATORY MANUAL

CHRISTOPHER GREEN | KRISTA CLARK | KAREN MATHIS | SUE BARKHURST | JENNIFER MANSFIELD

Kendall Hunt
publishing company

Cover image © 2015, Shutterstock, Inc.

www.kendallhunt.com
Send all inquiries to:
4050 Westmark Drive
Dubuque, IA 52004-1840

contents

laboratory safety rules

DRESS CODE

1. To protect yourself from possible injury, wear safety goggles whenever working with chemicals, burners, or any substance that may damage eyes. DO NOT wear contacts in the laboratory.
2. A lab apron or coat is recommended whenever working with chemicals or heated substances.
3. Tie back long hair when working with chemicals, burners, or other lab equipment.
4. Remove or tie back any loose articles of clothing or jewelry including scarves and bulky shirts or jackets. Shirts should have tight-fitting long sleeves and pants/slacks should be ankle length. Skirts are not appropriate—add leggings underneath for full protection or wear ankle-length skirts. Bare midriffs and low-cut necklines are not safe in the lab and will not be allowed.
5. Do not attempt to wash disposable gloves. Change them when they are dirty, contaminated, or ripped. Dispose of used gloves properly.
6. Do not wear sandals or open-toed shoes in the laboratory. Shoes must have closed toes and heels.
7. Know the location and appropriate use for safety eye wash, safety shower, fire extinguisher, and fire blanket.

GENERAL RULES

1. Never "horse around" in the lab. Never leave your lab activity unattended.
2. Be sure you understand all procedures in any lab investigation and possible hazards associated with it.

3. Read ALL directions for an investigation several times, and follow directions EXACTLY as they are written. Ask questions if you are not sure how to proceed.
4. Never perform unauthorized experiments.
5. Never handle equipment unless you have specific permission.
6. If spills occur, notify your instructor immediately.
7. No food or drink is allowed in the lab. Never eat or taste anything in the laboratory. This includes drinks, candy, and gum as well as chemicals. Do not apply makeup in the lab.
8. Notify your instructor of any medical problems you may have, such as allergies or asthma. Be sure your instructor has your emergency contact information.
9. Keep your laboratory area clean. Store bags, packs, and purses in appropriate places and off the lab tables. Do not handle electronic devices, phones, or keys while working in the lab. Be sure to clean your area thoroughly after you finish your activity. Wipe down the counters, and put away all equipment in clean, cool, and dry condition. Wash your hands before leaving the lab area.

FIRST AID

1. Report all accidents, spills, or broken glassware and equipment, no matter how minor, to your instructor immediately.
2. Know location of safety equipment and their proper use.

HEATING AND FIRE SAFETY

1. Never use a heat source without wearing safety goggles.
2. Never heat a chemical you are not instructed to heat.
3. Maintain a clean, uncluttered work area.
4. Never reach across a flame. Do not leave a lighted burner or other heat source unattended. Turn it off before walking away or ensure your partner is attending the experiment.
5. Use a hot plate instead of an open flame when flammable material is present.
6. Point a test tube or bottle that is being heated away from you and others. Do not heat directly on the bottom in a fixed place. Move the test tube through the heat source to avoid hot spots.
7. Never heat a liquid in a closed container.
8. Never pick up a heated container barehanded.
9. Do not dispose of matches or other solids into the sink. Rinse matches with water before disposing in the trash. Follow chemical disposal directions for chemical solids.
10. Never look directly into a laser beam. Doing so can cause permanent eye damage.

USING CHEMICALS SAFELY

1. Never mix chemicals for the "fun of it." Conduct only assigned experiments and only when the instructor is present.
2. Never touch, taste, or smell a chemical unless instructed to do so by your instructor. Keep your hands away from your face when working with chemicals.
3. If instructed to note the fumes in an investigation, waft the fumes away from the container toward your nose. Do not inhale fumes directly from the container.
4. If fumes are potentially dangerous, conduct procedure in a well-ventilated fume hood.
5. Notify your instructor IMMEDIATELY if chemicals are spilled.
6. Dispose of all chemicals as directed by your instructor.
7. Never return chemicals to their original containers. Never put spatulas or pipettes directly into reagent containers. Dispense some into a chemical dish and then use the quantity required. Dispose of excess chemicals properly.
8. Use extra caution when working with acids or bases. Pour over a sink or container, not over the lab table.
9. When diluting acids, ALWAYS pour acid into water to dissipate the heat produced. NEVER pour water into a concentrated acid.
10. Become familiar with safety precautions for each chemical to be used in an experiment. Know where eye-wash stations and fire safety equipment are located, as well as proper use.

USING GLASSWARE

1. Never force glass tubing into a rubber stopper. Use a lubricant such as glycerin to make the glass slide in easier.
2. Never heat glassware that is not thoroughly dry.
3. Use a wire screen to protect glassware from any flame.
4. Test glassware to be sure it is not hot before picking it up.
5. Never use broken or chipped glassware. If glassware breaks, notify your instructor and dispose of the glassware in the proper trash container.
6. Never eat or drink from laboratory glassware. Do not eat or drink in the laboratory.
7. Clean glassware thoroughly before putting it away.

USING SHARP INSTRUMENTS

1. Never cut material toward you; cut away from you.
2. Notify your instructor immediately if you cut yourself or receive a cut.

USING COMMON SENSE

1. Use common sense. This is not a complete list of all safety instructions but is intended as a guide. Never jeopardize yourself, anyone around you, or the lab facilities with your actions or inactions.
2. If in doubt, ask your instructor.
3. Eating and drinking in the labs are not allowed.

LAB ATTIRE

Please be aware that you will NOT be allowed to attend lab unless you are wearing long pants (leggings if wearing a dress) and you have on shoes that cover the ENTIRE top of your feet and the heel. Shoes can NOT have any holes in them, which means NO flip-flops, crocs, slippers, etc. Also be aware that we may be using Bunsen burners almost every week. Plan on pulling back long hair. You may want to limit clothing that is excessively loose or dangling. We will also be using permanent stains, so you may want to wear old clothing during those labs.

LABORATORY SAFETY RULES CONTRACT

I, the undersigned, having read the laboratory safety rules, statements of food and drink and lab attire, do agree to be in compliance with the mandates of the Biology Faculty/Staff as adopted on 08/22/2013. Failure to comply will result in non-acceptance into the class for the day (and loss of participation points and any other missed points) in which the student is in violation if the issue cannot be corrected within 10 minutes of class time. Quizzes may also be given over any or all of the safety information.

Personal and class safety are imperative and are of the highest regard in each class.

Student Signature: ____

Date: ____

hikes and mapping

Whenever you take a hike, you are required to document the experience.

All hikes must have observations (what you saw, where you saw it, weather/ground conditions, etc.) and a map with a key (including the campus buildings, a compass, beginning and end points, direction of hike, points of observations, etc.). Instead of collecting samples, photographs of plants, animals, landmarks, etc., will be added. You are required to have a minimum of three (3) photographs for each hike taken. All photographs must have a title and labels associated with them.

After each hike, you will write up a narrative summary of the hike, including plants and/or animals that you observed. Because the timing of hikes will vary for each class, there are pages located at the back of the book called "Hike Pages" that are to be used to record your information.

Each hike will have its own map, observations, photographs, and summary.

lab 1

Lab Techniques and Distinguishing between Precision and Accuracy

OBJECTIVES

After this lab, students will know how to:
1. Interpolate measurements
2. Correctly read a meniscus due to hydrogen bonding of water
3. Use and read a double pan balance
4. Distinguish the best uses of the three pieces of glassware being tested
5. Determine which piece of glassware is the most precise and the most accurate
6. Calculate statistical analysis of the means and the mean deviations of groups of data
7. Use bar graphs to display the mean deviations and the means of the three pieces of glassware

HYPOTHESIS

____

____

____

____

MATERIALS

- Double beam balance
- 400 mL beaker
- 250 mL beaker
- 100 mL graduated cylinder
- 100 mL volumetric flask;
- Dropping pipette
- ~ 150 mL dH_2O in the 400 mL beaker from the carboy
- Calculators

PROCEDURE

1. Obtain the 250 mL beaker, 100 mL graduated cylinder, and the 100 mL volumetric flask and draw each of them on the same page of your notebook, large enough to see the markings detail and label each piece of glassware with its respective name, at the bottom of each piece in the drawings.
2. Weigh each of the above named pieces of glassware DRY weight before water is added, record to 2 decimal places.
3. Fill out Table 1 with dry weight of the three pieces of glassware.
4. You will be weighing or massing out 100 mL of water in each glassware.
5. From the 400 mL reservoir with 150–175 mL of dH_2O, pour it into the 250 mL beaker and bring it to the 100 mL line on the beaker. You may use the dropping pipette for accuracy. Weigh the beaker plus water. Record as Beaker 1. Pour the water back into the reservoir 400 mL beaker. Repeat again, pour water into the 100 mL beaker and bring to 100 mL line. Reweigh beaker 2 and record.
6. Repeat the same procedures as in step 5 except using the 100 mL graduated cylinder. Read meniscus.
7. Repeat the same procedures as in step 5 except using the 100 mL volumetric flask. The line etched at the top is the 100 mL marker. Read meniscus.

Clean-Up Procedure

8. Pour the dH_2O down the drain and place each piece of glassware in a drying rack turned upside down.
9. Move markers on balance all the way to right of scale.
10. Return the balances to their proper storage.
11. Wipe up any water spills.
12. Return other equipment such as dropping pipettes to also drain and dry.

Statistical Analysis

13. For each piece of glassware, in each table, subtract out the dry weight of the piece of glassware, add the two numbers from the two trials together and divide by 2—this is the mean for that particular piece of glassware. Record this in the Average/Mean column. Do the same for the other two pieces

of glassware and record. This number will be used to determine the accuracy of the piece of glassware since we know that 100 mL of water should weigh approximately 100 g.

14. For each piece of glassware take the mean and subtract it from the difference between each of the two readings and divide by 2—this is the mean deviation and represents the amount of precision obtained by each piece of glassware. Precision is how close your set of numbers are to each other—that is, the repeatability factor.

15. Make two bar graphs in Excel, one for the determination of accuracy (mean) and one for the determination of precision (mean deviation). For the X axis use the type of glassware, and for the Y axis use the statistical numbers.

RESULTS

Table 1

	Beaker	Graduated Cylinder	Volumetric Flask
Dry weight			
dH_2O - Dry weight (reading 1)			
dH_2O - Dry weight (reading 1)			
Mean			
Standard deviation			

CALCULATIONS

CONCLUSIONS

Name ____ Date _____________________

1. Which piece of glassware is the most accurate according to your data?

2. Which piece of glassware is most accurate according to class data?

3. Which piece of glassware is the most precise according to your data?

4. Which piece of glassware is most precise according to class data?

5. Why you think the results came out as they did.

6. What is the best use for each of the pieces of glassware in the laboratory?

7. How did you like this lab? What suggestions do you have for next time this lab is taught?

Reference

Campbell and Reese. (2014). *Biology* (9th ed.). San Francisco: Pearson/Benjamin Cummings Publisher.

GENERAL BIOLOGY LABORATORY MANUAL

lab 2

Measurements Lab

After this lab, students will know how to:
1. Use different metric measurements in comparison to the English measurements
2. Switch between the two types of measurements
3. Use the various prefixes and their meanings with regard to the metric units
4. Experiment with various types of units of mass, volume, and length and be able to use conversion factors
5. Identify the various types of laboratory glassware and instruments
6. Properly use scientific notation

REFERENCE GUIDE

- Length
  - Meter (m) is the standard measure in metric
  - 1 m = 100 centimeters (cm)
  - 1 m = 39.4 inches
  - 2.54 cm = 1 inch
  - 1000 millimeters (mm) = 1 m
  - 10 mm = 1 centimeter (cm)

- 1000 m = 1 kilometer (km)
    - 1 km = 0.6214 mile (mi)
- Volume
    - Liter (L) is the standard measure in metric
    - 1 L = 1000 mL
    - 1 L = 1.06 quart (qt)
    - 946 ml = 1 qt
    - 1000 ml = 1 L
    - 1000 L = 1 m^3
    - 1 mL = 1 cm^3
- Mass
    - Gram (g) is the standard measure in metric
    - 1 kilogram (kg) = 1000 g
    - 1 kg = 2.2 pounds (lb)
    - 454 g = 1 lb
    - 1000 milligrams (mg) = 1 g
    - Weight is measure of gravitational pull on an object, but mass will always stay the same amount of matter.
- Time
    - Standard is second(s)
- Temperature
    - Standard is degrees Celsius (°C)
    - °C = [degrees F (°F) − 32] divided by 1.8
    - °F = 1.8(degrees C) + 32
- Prefixes
    - Kilo (k) = 10^3
    - Mega (M) = 10^6
    - Giga = 10^9
    - Deci (d) = 10^{-1}
    - Centi (c) = 10^{-2}
    - Milli (m) = 10^{-3}
    - Micro (µ) = 10^{-6}
    - Nano (n) = 10^{-9}
    - Angstrom = 10^{-10}
- Scientific Notation
    - Width of a human hair = 0.000008 m
    - In scientific notation 8×10^{-6} m
    - Hairs on a human scalp = 100 000
    - In scientific notation 1×10^5 hairs

NOTES

MATERIALS

- Meter and yard sticks
- Metric and foot rulers
- Various types of glassware
- Pan balances with weights
- Electronic balances
- Weighing boats
- Materials to measure for mass and length
- Containers to measure for volume
- Thermometers
- Stopwatch
- Various pieces of matter to measure

PROCEDURE/QUESTIONS

1. You may work alone or with a partner who is sitting next to you for this lab.
2. Select at least six substances to measure length, volume, and mass. Fill out Table 2 for each item for the three measurements. Record your metric measurements for each category and for each of the six items.
3. Without doing a lot of math, what is the best way to measure volume of a solid in a laboratory?

4. In completing the measurements, convert from metric units to English units, and state the measurements both ways. Also include the measurements in proper scientific notation when applicable.
5. For the "massing" of each substance first use the double pan balances with weights to determine the mass. Then use the electronic balances and compare the correlation of the data. (Some substances may not be within the range of the electronic balances.) Spend some time working with the double pan balances to problem solve how to do that when the actual beam scale only goes up to 10 g.
6. Work out this example: "Simply Lemonade" is said to be 11% lemon juice. A serving size is 13.5 oz; how many milliliters is that? In this amount of mL volume, according to the bottle's label, it contains 47 g of sugar (sucrose). Mass out (weigh is what we usually say but remember it is affected by the pull of gravity) 47 g of sugar and display this amount in a test tube.

7. Mass out 10 g of decorator sand in a weighing boat; be sure to "Tare out" the weighing boat or weight or subtract out the weight (mass) of the weighing boat from the boat with the decorator sand.

8. Place a thermometer into an empty beaker and let it equilibrate for at least 15 minutes. Record the temperature in °C and convert to °F. What is the temperature of the room in °C and in °F?

9. Using the stopwatches, record the amount of time it takes you to measure the exact length of your individual table in meters. What is the length of the individual lab tables in meters and centimeters? In yards and in feet?

10. Design and complete your own experiment involving measurement using the goals at the beginning of this lab as a reference point.

11. According to the website "How Stuff Works," a 2009 Toyota Camry weighs 3483 lb (gross vehicle weight) and a 2009 Ford Edge weighs 5300 lb. Respectively, how many kilograms would the Toyota weigh and how many kilograms would the Ford weigh? Discuss any hypotheses you might make concerning the weight of the car and gas mileage. Is there necessarily a direct correlation?

12. Write these numbers in proper scientific notation:
    a. .0000000095
    b. 150000000000

CALCULATIONS

RESULTS

Table 2

Item	Length		Volume		Mass	
	Metric	English	Metric	English	Metric	English
1.						
2.						
3.						
4.						
5.						
6.						

References

Campbell and Reese. (2014). *Biology* (9th ed.). San Francisco: Pearson/Benjamin Cummings Publisher
Timberlake, Karen. (2010). *General, Organic and Biological Chemistry Structures of Life* (3rd ed.). New York: Prentice Hall.

lab 3

Determining the pH of Common Substances

INTRODUCTION

The pH of a substance is an indication of how acidic or how basic the substance is. In terms of chemistry, the pH is the potential of hydrogen in a solution. Hydrogen ions in solution can produce a charge. This charge can be detected using a pH meter or pH paper. The pH scale ranges from 1–14 and the lower the number the more acidic (the more hydrogen ions present) a solution is. In distilled water the pH is 7. A one number difference in pH represents a 10-fold difference in acidity.

Another way to define an acid is to say an acid donates a H^+ (hydrogen ion) to a solution and a base accepts a H^+.

OBJECTIVES

After this lab, students will know how to:
1. Test for pH using a pH meter and pH paper
2. State the pH of common substances

HYPOTHESIS

____

____

____

____

____

____

____

MATERIALS

- pH meter
- pH paper
- Various test substances

PROCEDURE

1. The pH meter you will be using is a handheld instrument.
2. When not in use, the pH meter must be kept in buffer so the electrode will not be damaged.
3. When taking a pH measurement be sure to rinse the buffer off the electrode with distilled water and then insert the tip of the electrode into the solution to be tested.
4. Then take and record the digital reading.
5. Rinse the substance off the electrode with distilled water and then return the electrode to the buffer solution; thus, preventing contamination into the buffer.
6. To test with pH paper, remove a small strip of test paper and dip it in the substance. Compare the color with the chart and record the pH.
7. Now, determine the pH of your two substances and record them in Table 3. As time permits, determine the pH of other substances. You may consult your classmates to complete your data table.

8. Create a graph of the pH of common substances tested in class. This graph can be made using Excel or by hand. The X axis should contain the substances and a label. The Y axis should be the pH scale and a label. The graph should also have a title. A column graph is appropriate.

RESULTS

Substance	pH (meter)	pH (paper)	Acid, Base, or Neutral

CONCLUSIONS

Name ____ Date _____________________

QUESTIONS

1. Based on the substances tested, make some general statements about type of substance and pH value.

2. The pH meter is more accurate than the pH paper. What are some circumstances when pH paper alone would be sufficient to use?

3. If one substance has a pH of 5 and another a pH of 3, how many times more acidic is the second substance?

4. Based on your textbook reading, describe what a buffer will do to an acid? to a base?

5. At the drug store you can buy buffered aspirin. The scientific name for aspirin is acetylsalicylic acid. What properties would the buffered aspirin have over normal aspirin?

6. Describe each chemical below as either an acid or a base.
    a. HCl
    b. H_2SO_4
    c. NaOH
    d. KOH

Reference

Campbell and Reese. (2014). *Biology* (9th ed.). San Francisco: Pearson/Benjamin Cummings Publisher.

lab 4

Biological Indicators

INTRODUCTION

Carbohydrates that are relatively small are called sugars and very large carbohydrate molecules are called polysaccharides. Sucrose is an example of a sugar and starch is an example of a polysaccharide. Chemicals can be used to detect the presence of sugar and/or starch in a solution. When a chemical changes color to detect the presence of another molecule, that chemical is known as an indicator. Iodine, bromothymol blue, and Benedict's reagent are all indicators.

OBJECTIVES

After this lab, students will know how to:
1. Determine which of the three indicators, if any, will detect sugar
2. Determine which of the three indicators, if any, will detect starch

NOTES

HYPOTHESIS

____

____

____

____

MATERIALS

- Test tubes
- Graduated cylinders
- Beakers
- Test tube rack
- Droppers
- Hot water bath
- Sugar and starch solutions

PROCEDURE

1. As an indicator, iodine turns from yellow to bluish purple. Five drops is adequate for every 5 mL of test solution.
2. As an indicator, bromothymol blue turns from blue to yellow. Ten drops is adequate for every 5 mL of test solution.
3. As an indicator, Benedict's reagent turns from blue to rusty yellow/orange. Benedict's reagent must be heated with the test solution in order to work as an indicator. Ten drops is adequate for every 5 mL of test solution.
4. You must set up an experiment to test the indicators and find out their role. You have access to the following: test tubes, graduated cylinders, beakers, test tube rack, droppers, hot water bath, and separate solutions of sugar and starch. If you need additional items, ask the facilitator.
5. Be sure to record the following:
   a. Your team's prediction (hypothesis) as to which indicators will detect the presence of sugar and/or starch
   b. Your procedure
   c. Your results
   d. Your conclusions

PROCEDURE

RESULTS

CONCLUSIONS

Name ____ Date _______________________

1. Was your team's hypothesis for the indicator lab correct? Why or why not?

2. What were your conclusions for the indicator experiment for each indicator tested?

Reference

Campbell and Reese. (2014). *Biology* (9th ed.). San Francisco: Pearson/Benjamin Cummings Publisher.

lab 5

Testing for Lipids in Foods

INTRODUCTION

Lipids are one type of macromolecule (macro = large) and are important components of cells. Lipids are divided into three groups: fats, phospholipids, and steroids. Fats are long chains of hydrocarbons attached to a glycerol molecule. Fats store energy in the cell. Phospholipids are molecules that make up a large part of the cell membrane. They have one end that is hydrophilic (water loving) and another end that is hydrophobic (water fearing). Steroids are structurally different than the other two types of lipids (i.e., they are composed of fused ring structures) and are important components in many hormones.

In nutritional terms, the lipids are most often discussed as fats. Fats are divided into saturated and unsaturated fats. Saturated fats do not have any double bonds in their fatty acid tails (hence they are "saturated" with hydrogen) and are fats found from animal sources (e.g., lard, bacon, fat around meat). Unsaturated fats have one or more double bonds in their fatty acid tails and are derived from plant sources (e.g., corn oil, olive oil, sunflower oil).

OBJECTIVES

After this lab, students will know how to:
1. Extract lipids from a food source
2. Identify lipids as saturated or unsaturated
3. Calculate percent error

- Chocolate chips (semisweet)
- Sunflower seeds (no shells)
- Potato chips
- Acetone microwave
- Balance or scale
- Hammer
- Paper towels
- 250 mL beaker
- Foil
- 100 mm Petri dishes
- Latex or rubber gloves
- 400 mL beakers
- Safety goggles
- 50 mL graduated cylinder
- Glass stir rod
- Sharpie or marking pencil

PROCEDURE

Part A: Visual Inspection of Lipids

Team 1. Chocolate Chips
1. Measure out 2 g of chocolate chips and place on a paper towel.
2. Microwave for 40 seconds on high.
3. Fold the paper towel over the chocolate chips and gently press the chocolate chips flat with your fingers.
4. Allow it to sit for 5 minutes. Open up the paper towel. Record your results in Table 4.

Team 2. Potato Chips
1. Measure out 2 g of potato chips and place on a paper towel.
2. Microwave for 25 seconds on high.
3. Fold the paper towel over the potato chips and crush the chips with a beaker.
4. Allow it to sit for 5 minutes. Open up the paper towel. Record your results in Table 4.

Team 3. Sunflower Seeds
1. Measure out 2 g of sunflower seeds and place on a paper towel.
2. Microwave for 25 seconds on high.
3. Fold the paper towel over the sunflower seeds and crush the seeds with a beaker.
4. Allow it to sit for 5 minutes. Open up the paper towel. Record your results in Table 4.

Part B: Quantitative Measurement of Fats

Team 1. Extraction of Fat from Chocolate Chips
1. Weigh out 5 g of chocolate chips (9 chips). Crush the chocolate between two sheets of foil with a hammer.
2. Label the beaker and the Petri dish that your team will be using.
3. Weigh the empty beaker and record this weight in your notebook.
4. Making sure to zero out the scale once the empty beaker is on, record the weight with the food inside. Record in Table 5.
5. Add 10 mL of acetone to the food in the beaker.
6. Swirl for 1 minute in a hood, or stir with a glass rod (in a well-ventilated area).
7. Carefully decant (drain off the liquid from) the acetone into the Petri dish, making sure the chocolate remains in the beaker.
8. Add 10 mL of acetone to the food again and repeat steps 5 and 6.
9. Allow the acetone in the Petri dish to dry overnight or until next class period in a hood (or a well-ventilated area) to visualize the lipid that was extracted.
10. Allow the beaker with the food to dry overnight or until next class. Weigh the beaker with the food.
11. Describe the fat left in the Petri dish. (Is it solid or liquid?) Record your observations.

Team 2. Extraction of Fat from Potato Chips
1. Weigh out 5 g of potato chips. Break into dime-size pieces with your fingers.
2. Repeat steps 2–11 listed for Team 1.

Team 3. Extraction of Fat from Sunflower Seeds
1. Weigh out 5 g of sunflower seeds. Crush the seeds between two pieces of foil with a hammer.
2. Repeat steps 2–11 listed for Team 1.

OBSERVATIONS

____

____

____

____

____

____

RESULTS

Weight of Empty Beaker ________________

Table 4

Food	What you saw on the paper towel
Chocolate chips	
Potato chips	
Sunflower seeds	

Formulas for Table 5

(Weight of Food and Beaker after 24 Hours) – (Weight of Empty Beaker) = Weight of Food after 24 Hours

(Weight of Food) – (Weight of Food after 24 Hours) = Weight Lost from Food Overnight

(Weight Lost from Food Overnight) / (Weight of Food) x 100 = Percent Lipid Extraction

$$\frac{(\text{Percent Lipid Extraction}) - (\text{Percent Lipid from Nutrition Label})}{\text{Percent Lipid from Nutrition Label}} \times 100\% = \text{Percent Error}$$

Table 5

Food	Weight of Food (g)	Weight of Food and Beaker after 24 Hours (g)	Weight Lost from Food Overnight (g)	Percent Lipid Extraction (%)	Percent Lipid from Nutrition Label (%)	Percent Error (%)
Chocolate chips						
Potato chips						
Sunflower seeds						

QUESTIONS

1. How do you know the spot on the paper towel is fat and not water?

2. Determine which lipids contained saturated and unsaturated fatty acids in this experiment, based on your descriptions of the fats in the Petri dishes.

3. Which food had the most percent error?

4. Which had the least?

5. What are potential sources of error in this procedure and why are there differences between the foods?

References

Adapted from Institute of Food Technologists, IFT Experiments in Food Science Series.
Campbell and Reese. (2014). *Biology* (9th ed.). San Francisco: Pearson/Benjamin Cummings Publisher.

lab 6

Cells

INTRODUCTION

This laboratory will give you the opportunity to examine various kinds of cells under the microscope. You will be given a variety of materials and look for visible organelles. For each sample, focus and examine at each magnification, but **draw** and **label** a representative cell at 400x total magnification (TM). Take your time and try to be as accurate as possible with your drawings.

MATERIALS

- Light microscope
- Disposable slides
- Onion specimen
- Plant specimen
- Potato specimen
- Toothpicks

NOTES

PROCEDURE/ QUESTIONS

1. Draw and label each of the following cells at 400x TM.
   a. Cheek cell or buccal smear. Your teacher will demonstrate how to make a buccal smear. All students should make their own buccal smear and examine it under the microscope. Draw and label any organelles that are visible. (The small dark "dots" scattered around the cells are bacteria from the inside of your mouth.) *Note: you should see at least two cell parts clearly.*

   b. Take a very thin layer of onionskin and create a wet mount. Examine the onion cells under the microscope. Draw and label any visible organelles. Note how the shape is different from the cheek cells. *Note: you should see at least two cell parts clearly.*
   c. Pick a green leaf from a tree outside. Get a very thin section and examine it under the microscope. Can you see **chloroplasts**? You may be able to see tiny openings or pores (especially on the underside of the leaf). These are called stomata and allow gases to enter and leave the leaf during photosynthesis.
   d. Examine the plant sample provided in class and look for chloroplasts. Draw and label what you see. Can you see the chloroplasts moving in the cell?
   e. Take a thin slice of potato and make wet mount using iodine as the liquid. Look for organelles within the potato cells that turn purple (purple dots). What substance must be contained within these organelles? Hint: What is iodine and indicator for and what type of nutrient is found in potatoes? The organelle you are observing is called an *amyloplast* (from the root words *amyl* = starch and *plastid* = storage organelle).
2. Which of the cells you observed were the largest?

3. Based on the organelles you observed, complete Table 6. You should include the following: nucleus, cell membrane, chloroplast, amyloplast, cell wall.

4. Which cell was the most interesting for you to observe? Why?

5. Name three differences between plant and animal cells in terms of structure/organelles.

RESULTS

Organelle	Shape/ color/ describe	Specimens you observed w/ organelle	Function
Nucleus			
Cell membrane			
Chloroplast			
Amyloplast			
Cell wall			

Figure 1 Buccal smear 400x TM

Figure 2 Onion 400x TM

Figure 3 Green leaf 400x TM

Figure 4 Plant 400x TM

Figure 5 Potato 400x TM

Reference

Campbell and Reese. (2014). *Biology* (9th ed.). San Francisco: Pearson/Benjamin Cummings Publisher.

lab 7

Yeast Fermentation

OBJECTIVE

1. After this lab, students will know how to observe the effect of environment (sugar levels) on yeast activities.

HYPOTHESIS

MATERIALS

- Active dry yeast
- ruler
- String
- Flasks and balloons (size of flask determined by size of balloon)
- 250 mL beaker
- Sugar and measuring spoons
- Warm water (about 110°F)

PROCEDURE

1. Each group will create three replicates of their assigned conditions.
2. Measure 100 mL of warm water and pour into the flask.
3. Add ½ tsp of yeast to the flask and stir until it is in solution.
4. Add the assigned amount of sugar to each flask.
   a. Group 1: After pouring your yeast solution into the flask, immediately place the balloon on the top of the flask. Note the time.
   b. Group 2: After pouring your yeast solution into the bottle, add ¼ tsp of sugar to the mixture and swirl it gently until it dissolves. Then place the balloon on top of the bottle. Note the time.
   c. Group 3: After pouring your yeast solution into the bottle, add ½ tsp of sugar to the mixture and swirl it gently until it dissolves. Then place the balloon on top of the bottle. Note the time.
   d. Group 4: After pouring your yeast solution into the bottle, add ¾ tsp of sugar to the mixture and swirl it gently until it dissolves. Then place the balloon on top of the bottle. Note the time.
   e. Group 5: After pouring your yeast solution into the bottle, add 1 tsp of sugar to the mixture and swirl it gently until it dissolves. Then place the balloon on top of the bottle. Note the time.
5. Using your string, measure the size of your balloon every 5 minutes for 45 minutes. (A final measurement 12 or 24 hours later may also be made.)
6. Record this information in Table 7.
7. Determine the change in size of balloon diameter and calculate the average change in diameter at each measurement. Record these on a data table and collect data from the entire class.

RESULTS

Table 7

Time (min)	Diameter (cm)				Change
	1st measurement	2nd measurement	3rd measurement	Average	
5					
10					
15					
20					
25					
30					
35					
40					
45					
12 hours					
24 hours					

CALCULATIONS

CONCLUSIONS

Name ____ Date _____________________

QUESTIONS

1. Which of the balloons reached the largest size?

2. What material was collected in the balloons?

3. What were the variables in this experiment?

4. Which one of the groups was the control? Why was it the control?

5. What is fermentation? What are the products of fermentation?

6. What caused the balloons to increase in size?

7. Name two other variables that may affect fermentation rates and could be tested.

8. Yeast is used in making bread. Using the information obtained in this experiment, explain what causes bread to rise and what conditions are necessary for optimum rising?

9. Make a multiline graph of the change in balloon diameter over time in MS Excel. Include a legend.

Reference

Campbell and Reese. (2014). *Biology* (9th ed.). San Francisco: Pearson/Benjamin Cummings Publisher.

lab 8

Introduction to Chromatography

INTRODUCTION

Chromatography is the process by which a mixture of substances is separated into individual components. There are several different chromatography methods and one of the simplest is paper chromatography. There are two phases in this method: stationary phase and mobile phase. The stationary phase is the paper and the mobile phase is the solvent (petroleum ether). As the solvent moves over the sample, molecules will have varying degrees of solubility in the solvent. The most soluble molecules will travel the furthest.

Paper chromatography will be used to separate the pigments that are found in green plants. Green plants contain several pigments (not just the green chlorophyll we generally observe). These are most visible in the fall when the chlorophyll degrades and the yellows, oranges, and reds are visible.

OBJECTIVE

After this lab, students will know how to use paper chromatography and observe the pigments obtained in spinach extract.

MATERIALS

- Spinach extract
- Solvent
- Chromatography paper

PROCEDURE

1. Your instructor will demonstrate the proper use of paper chromatography. Follow the method presented in class. In brief, a spot of extract will be applied to the chromatography paper. The edge of the paper will be placed in the solvent. As the solvent moves up the paper, various pigments will be observed.
2. As soon as the solvent reaches the top of the paper, make a drawing (life size) of the pigments and indicate their location and color (Figure 6).

Name ____ Date _____________________________

1. How many pigments were observed in the sample?

2. The color of pigments means that those wavelengths of light are not absorbed, but rather reflected back to us. Based on your results, what colors of light are being *absorbed* by the plant?

3. The petroleum ether is nonpolar (not water soluble). When red cabbage extract is placed on chromatography paper, a red dot remains at the same place it was applied. What does this indicate about the pigment in red cabbage?

4. Describe another experiment or question that could be answered about plant pigments using paper chromatography.

Figure 6 Chromatography paper results

Reference

Campbell and Reese. (2014). *Biology* (9th ed.). San Francisco: Pearson/Benjamin Cummings Publisher.

lab 9

DNA Extraction

INTRODUCTION

DNA is the molecule of life. Later on in the semester you will take a detailed look at the structure of DNA, and how DNA codes for traits. Today, you are going to extract some DNA from wheat seeds.

DNA is found in the nucleus of each cell. During extraction several things must happen to allow the DNA to leave the nucleus. First, the outer wall of the plant cells (the cell wall) must be broken. Next, several proteins will be removed to allow us to see the DNA. Finally, the DNA will precipitate into alcohol where it is not soluble.

First, several solutions are prepared and the wheat germ ground up. The wheat germ is then placed into a detergent solution to break up the cells further. Heating to 60°C will break down some proteins but not the DNA (DNA will break down at 80°C).

OBJECTIVES

After this lab, students will know how to extract DNA from cells.

MATERIALS

- Wheat germ
- Detergent
- Cheese cloth
- Test tube
- Spooling rod

PROCEDURE

1. You may work in groups of two or three students.
2. Add 1 tsp of wheat germ (untoasted) to a beaker and add 20 mL of hot (50°–60°C) tap water. Mix constantly for 3 minutes.
3. Add 1 mL of detergent and mix gently for 5 minutes. Try not to create any foam.
4. Filter the mixture through cheese cloth and then add the filtrate to a test tube.
5. Lower the spooling rod into the solution and let it rest on the bottom of the test tube.
6. Angle the test tube and SLOWLY add 4 mL of cold alcohol to the solution. You want to create two layers (alcohol on top and water on the bottom).
7. Rotate the spooling rod clockwise without touching the sides of the test tube. Masses of DNA should stick to the rod at the alcohol/water interface.

OBSERVATIONS

Name ____ Date _________________________

1. Describe the appearance of the extracted DNA.

2. Why does the DNA appear at the water/alcohol interface?

3. What is the function of DNA in a cell?

4. What additional experiments could be performed based on this procedure?

Reference

Campbell and Reese. (2014). *Biology* (9th ed.). San Francisco: Pearson/Benjamin Cummings Publisher.

lab 10

DNA Structure and Replication

Watson and Crick discovered the structure of DNA in 1953 and were awarded the Noble Prize for their efforts. Their publication was about one page in length. Rosalind Franklin was instrumental in discovering the helical shape of DNA, but died before the Nobel Prize was awarded. *The Double Helix* is a book written from Watson's perspective about the discovery.

DNA replication takes place before all cell division—that is, mitosis and meiosis. DNA replication occurs when DNA "unzips" and then two new strands are formed opposite the first two strands. This is called semiconservative replication, because the new molecule has one old and one new strand. Chemicals in the cell (enzymes called DNA polymerase) perform a proofreading function to ensure all the base pairs match properly and can repair DNA.

When the molecule unzips, nucleotides can only be added in a certain direction: from 5' (phosphate end) to 3' (sugar end). So the strand that ends in a 3' or sugar has continuous addition of the nucleotides. This is called the leading strand.

Things such as radiation, UV rays, and X-rays can alter and damage (mutate) DNA. Sometimes our bodies can repair the damage. Other times disease (e.g., some kinds of skin cancer) will result if the repair mechanism is damaged.

These understandings have led to the ability to figure out the nucleotide sequence in DNA. Polymerase chain reaction has allowed many copies of DNA to be made. (This is why only a small sample of DNA is needed at a crime scene.)

OBJECTIVES

After this lab, students will be able to construct DNA using model kits, as well as using those same kits to model DNA replication.

MATERIALS

- DNA model kit

PROCEDURE

Part 1: DNA Structure

1. Assemble all the nucleotides in your kit. Notice the base changes, but is only one of four colors.
2. In DNA, the sugar and phosphate for the backbone of the molecule and the bases are in the center. Assemble one strand in the following order
   a. A C T A C G A G T C T T G G
   b. Adenine is orange, Guanine is green, Cytosine is blue, and Thymine is yellow.
3. But DNA is a double-stranded molecule. Once again the sugar and phosphate are the backbone; and the types of bases that end up opposite one another are defined. A always bonds with T and G always bonds with C. A bond called a hydrogen bond joins the two bases together.
4. Assemble the second strand of the DNA molecule according the base pair rules. Use hydrogen bonds to connect the bases. Now hold your DNA molecule up and twist it slightly.
5. DNA is called a double helix. Notice:
   a. sugar phosphate backbone
   b. bases in center
   c. slight twisted shape of molecule
   d. Also note that the two strands are offset slightly. The end that has the phosphate is called the 5' end. The end that has the sugar is called the 3' end. The DNA molecule is called antiparallel because the strands are positioned in opposite directions of each other.
6. Use your completed strand to fill out Table 8.

Part 2: DNA Replication

7. Practice replicating the DNA. Pull apart the two strands and reconstruct a corresponding strand for each original strand.
8. Your two new molecules of DNA should look exactly alike. Compare them. Write your observations below.

RESULTS

Table 8

Feature of DNA	Description
Name of sugar in DNA	
Nitrogen bases found in DNA	
How many strands make up the molecule?	

QUESTIONS

1. Describe how DNA replicates.

2. What happens to the original parent strands when DNA replicates?

3. The following is a simplified version of DNA. Indicate the base pairs for strand 2.

   Strand 1 TACGCCGGTAACTACG

   Strand 2

Reference

Campbell and Reese. (2014). *Biology* (9th ed.). San Francisco: Pearson/Benjamin Cummings Publisher.

lab 11

Mitosis

INTRODUCTION

Mitosis is the process of cell division in which two daughter cells look exactly like the parent cell. The DNA in the parent cell replicates before mitosis begins. During mitosis, the nuclear envelope disintegrates and the chromosomes thicken (prophase). At metaphase, the chromosomes line up in the center of the cell. During anaphase the two copies of DNA (sister chromatids) separate and move to opposite sides of the cell. Telophase is like the opposite of prophase where the nuclear envelope reforms and the chromosomes decondense. The actual division of cytoplasm to produce two new cells is called cytokinesis. Often, telophase and cytokinesis occur at the same time. In animals, the division furrow results in a pinching of the cytoplasm to produce new cells. In plants, the cell plate forms in the middle of the two cells and serves as the template for the cell wall.

OBJECTIVE

After this lab, students will know how to identify and describe the stages of mitosis as well as interphase.

MATERIALS

- Onion root tip slide

PROCEDURE

1. One good place to study the process of mitosis is to examine a portion of the organism undergoing rapid growth, such as in the root tip of a plant.
2. The slide you have contains three samples of onion root tips. Mitosis will not be at the very tip of the root, but in the layer of cells just below the root cap.
3. Remember, roots and cells are three dimensional, so a cross section of the root will produce samples that may not have any nuclei visible or nuclei at various angles.
4. Look for clear examples of cells and find a cell that clearly shows each stage of mitosis. You must have a drawing for each of the following: Interphase, Prophase, Metaphase, Anaphase, and Telophase/Cytokinesis.
5. Use your textbook to help you identify the stages. In addition, label your drawings (including chromosomes, spindles, cell wall, nuclear envelope when visible).
6. Spend some time creating quality, representative drawings. Part of your grade for this laboratory will be based on the quality of your illustrations.

RESULTS

Figure 7 Root Tip Interphase 400x TM

Figure 8 Root Tip Prophase 400x TM

Figure 9 Root Tip Metaphase 400x TM

Figure 10 Root Tip Anaphase 400x TM

Figure 11 Root Tip Telophase/ Cytokinesis 400x TM

QUESTIONS

1. Why is mitosis necessary in living things?

2. What are two key differences between plant and animal cells undergoing mitosis?

3. What are sister chromatids? How are they formed?

4. Prokaryotic cells do not reproduce by mitosis. Explain how prokaryotic cells reproduce and give the name of this process.

Reference

Campbell and Reese. (2014). *Biology* (9th ed.). San Francisco: Pearson/Benjamin Cummings Publisher.

lab 12

Human Genetics

INTRODUCTION

Modern genetics began with Gregor Mendel's experiments with pea plants. Mendel hypothesized that there are alternative forms of genes (the units that determine heritable traits). The inheritance of many human traits follows Mendel's laws and are controlled by a single gene (e.g., freckles, widow's peak, or free earlobes). Many disorders also follow the same laws, such as albinism, cystic fibrosis, and sickle cell anemia.

OBJECTIVE

After this lab, students will know how to identify several human characteristics and be able to interpret both genotype and phenotype for each trait.

MATERIALS

- None

PROCEDURE

Several human traits are controlled by simple Mendelian genetics. For each trait listed in the table, determine your phenotype and possible genotypes. If you know only one allele in your genotype, use a dash (-) to represent the second, unknown allele. Your instructor will show examples of the traits.

Examples: I have a bent pinky (my phenotype) and so I know I have at least one dominant allele, P. My genotype could be PP or Pp. Since I do not know I will just write P-. My friend does **not** have a widow's peak (phenotype). This means we know for certain her genotype is ww.

Trait	Description	Phenotype	Genotype
Widow's peak	Peak is dominant (W)		
Thumb crossing	Left over right is dominant (F)		
Hair texture	Curly is dominant (W); wavy is incomplete dominance (Ww) and straight is recessive (ww)		
Hitchhiker's thumb	Straight thumb is dominant (S) over curved (ss)		
Bent little finger	Bent pinky is dominant (P) over straight pinky (pp)		
Hair on mid-digit	Use index finger; mid-digit hair is dominant (M) over no hair		
Dimples	Dimples are dominant (D) over no dimples (dd)		
Ear lobes	Free earlobes are dominant (F) over attached earlobes (ff)		
Tongue roller	Rolling the tongue is dominant (R) over non-rollers (rr)		

lab 13

Corn Genetics

INTRODUCTION

This exercise will demonstrate some of the phenotypic differences that can occur in corn plants. Two different sets of seeds were planted. Given the genotypes of the parents, you will make predictions for the outcome of the offspring. You will then compare predicted outcomes to actual experimental results.

OBJECTIVES

After this lab, students will know how to:
1. Complete Punnet squares to determine genotype and phenotype
2. Explain Mendel's first law of segregation
3. Explain Mendel's second law of independent assortment

HYPOTHESIS

____

____

____

____

MATERIALS

- Corn samples

PROCEDURE

Set 1

There is a recessive gene that does not allow the corn to produce chlorophyll and thus produces albino plants. If G = green plants and g = albino plants, what would the offspring of Gg x Gg look like? Complete the Punnet square in your notebook. Write this prediction down in Table 9 as the form of a percent. Now, go and count the number of offspring of this match and record those numbers in Table 9.

Table 9

Phenotypes	Predicted Percentage	Experimental Results	
		Number	**Percent**
Green			
Albino			
Total			

Set 2

A dihybrid cross of corn was made using the following traits: T = tall, t = dwarf and G = green, g = albino.

What are the four phenotypes possible? Record these in Table 10. What would the predicted offspring of the following cross be? Complete the Punnet square in your notebook.

TtGg x TtGg

Record these percentages in Table 10.
Now, go and count the number and phenotypes of offspring and record those numbers in Table 10.

Table 10

Phenotypes	Predicted Percentage	Experimental Results	
		Number	Percent
Totals			

CONCLUSIONS

____

____

____

____

Name ____ Date _______________________

QUESTIONS

1. For the green and albino corn, did the predicted results match the experimental results? Discuss the outcomes.

2. Would the albino plants survive to reproduce? Explain.

3. For the dihybrid cross, did the predicted results match the experimental results? Discuss the outcomes.

Reference

Campbell and Reese. (2014). *Biology* (9th ed.). San Francisco: Pearson/Benjamin Cummings Publisher.

lab 14

Fruit Fly Genetic Crosses (Virtual Lab)

The fruit fly (*Drosophila melanogaster*) is one of the most common animals used to study genetics. There are several observable traits that can be used in genetic crosses. In the laboratory, the flies are easily mated. They can be "knocked out" with ether and observed using a dissecting scope. This lab uses computer animation to simulate fruit fly genetics.

After this lab, students will know how to:
1. Explain the way fruit flies are used in genetics labs.
2. Make genetic crosses using fruit flies to demonstrate monohybrid and sex-linked genetics.

____

____

____

MATERIALS

- None

WEBSITE

http://www.sciencecourseware.com/vcise/
You can enter this site as a guest and copy data into your laboratory notebook.

PROCEDURE

1. Go to the website listed in the protocol and make the genetic crosses indicated below.
2. In each scenario you will cross the P (parental) generation and the resulting F_1 (filial) generation.
3. Based on the results you will determine the genotypes of each parent and also determine if the trait examined is sex linked.
4. Wing length
    a. Cross wild type (normal wing) females with vestigial (short wing) males.
    b. Fill in Table 11 with the results.
    c. Then, cross the F_1 generation and complete Table 11.
5. Eye color
    a. Cross the wild type (red eyed) female with the white eyed males.
    b. Fill in Table 12 with the results.
    c. Then, cross the F_1 generation and complete Table 12.

Table 11

Phenotypes	Predicted Percentages	Experimental Results	
		Number	Percent
Green			
Albino			
Total			

Phenotypes	Predicted Percentages	Experimental Results	
		Number	Percent
Totals			

CONCLUSIONS

____

____

____

____

____

____

QUESTIONS

1. Is wing length a sex-linked trait? Explain.

2. What are the genotypes for the parents in the P cross and the F_1 cross of the wing length trait? Draw a Punnet square for each cross.

3. Is eye color a sex-linked trait? Explain.

4. What are the genotypes of the parents for the P cross and the F_1 cross of the eye color trait? Draw a Punnet square for each cross.

5. Comment on the use of computer simulations for this activity as opposed to using actual fruit flies. Are you in favor of these types of exercises?

References

Campbell and Reese. (2014). *Biology* (9th ed.). San Francisco: Pearson/Benjamin Cummings Publisher. http://www.sciencecourseware.com/vcise/

lab 15

ABO-Rh Blood Typing Lab

(Used with Carolina Biological Supply Synthetic Blood Kit)

INTRODUCTION

There are several characteristics of blood, but the most common are the blood types (A, B, AB, and O) and Rh factor (positive or negative). Blood type can be determined because the antigens found on the red blood cells will react with certain antibodies in the plasma.   For example, a person with type A blood will have A antigens on the red blood cells, but his or her plasma will carry B antibodies. (If this person had A antibodies or was exposed to them, an antigen/antibody complex would form that would agglutinate the blood.) Although not helpful in the body, the reaction of antibody to antigen allows blood type testing to occur.

OBJECTIVES

After this lab, students will know how to:
- Explain the way blood typing is performed
- Explain the genetics of blood type

HYPOTHESIS

____

____

____

____

MATERIALS

- Synthetic Blood Kit

PROCEDURE

1. Using the dropper vial, place a drop of the first sample (#1) in each well of the blood typing slide. Replace the cap on the dropper vial (before opening another).
2. Add a drop of synthetic anti-A (blue) to the well labeled A. Replace the cap.
3. Add a drop of anti-B (yellow) to the well labeled B. Replace the cap.
4. Add a drop of anti-Rh (clear) to the well labeled Rh. Replace the cap.
5. Using a different color stick to match each well, gently stir the synthetic blood and the anti-serum. Avoid contamination.
6. Examine the thin films of liquid mixture left behind. If the film remains uniform in appearance, there is no agglutination. If the sample appears granular, agglutination has occurred. Indicate "yes" or "no" for agglutination in Table 13. A positive agglutination reaction indicates the blood type.
7. Thoroughly rinse the slide and repeat the procedure for samples 2–4.

Table 13

	Sample 1	Sample 2	Sample 3	Sample 4
Anti- A				
Anti-B				
Rh				
Blood type				

CONCLUSIONS

QUESTIONS

1. A woman with blood type B (ignore the Rh factor here) gives birth to a child who has blood type A. She claims the man with sample 4 blood is the father. Is this possible? Explain.

2. Of the four samples, who are potential fathers?

References

Campbell and Reese. (2014). *Biology* (9th ed.). San Francisco: Pearson/Benjamin Cummings Publisher.
Carolina Biological Supply Synthetic Blood Kit.

lab 16

Introduction to the Scientific Paper

INTRODUCTION

The scientific paper is the format used by scientists when they publish the results of their experiments. Scientific publications are peer reviewed, which means that the paper must past the scrutiny of other scientists in that area of research. Each paper that is submitted is read by two or three other scientists and the experiments, data, and conclusions are closely examined. The reviewers are looking for controlled experiments, proper data analysis, and sound conclusions. If the paper passes the review it will be published in a journal (a scientific magazine).

Regardless of the field (Botany, Medicine, Embryology, etc), all scientific papers will have the same format. The sample paper given to you will help you learn the parts of a scientific paper and the types of information found in each part.

Titles: The title of the scientific paper should be specific. Note the title of your sample article, "Effects of the emerald ash borer invasion on four species of birds."

Abstract: Every article must include an abstract. The abstract is generally a paragraph long and summarizes the entire experiment. The objective, methods, results, and conclusions are all included. The abstract is a good way to get an overview of a paper without reading the entire paper. Look for statements that summarize the objective, methods, results, and conclusions in the abstract of the sample paper.

Introduction: The introduction will introduce the reader to the topic, often explain why the study is important, and describe the objectives of the experiment or study. Find the paragraph that describes the objectives of the study.

Methods (sometimes called Materials and Methods): The materials and methods section only describes how the experiment was performed (written in past tense). This section only provides information about how the data were collected (not what those data were). Notice that there is not a list of materials used. This is different than your lab protocols that do list what is needed. A scientific paper never has a straight list of materials. The reader would have to read through the entire section to find out what materials were used. Often this section is subdivided.

Results: The results section of the paper summarizes the data collected and/or observations made. Even when figures (graphs) and tables are used to display the data, the results section contains paragraphs that describe the trends found in the data. Any figure or table included in the paper must be referenced in the text of the paper. No conclusions are made in this section. Only observations or measurements are included. Interpretations of the data occur in the next section.

Figures can be found in several sections of the paper, but most often are found in the Results section. Figures can contain diagrams, pictures, or graphs. Tables are not considered Figures, and are called Tables in a paper.
- Notice that each Figure has a complete Figure Label which is a complete sentence describing what is found in the figure.
- Notice that each Figure is referred to somewhere in the body (text) of the paper. A figure must be referred to in the text.

Discussion: In this section the results are interpreted and conclusions drawn. The author(s) try and make sense of the data. If the data are confusing the author(s) may try and explain the strange results. Often, flaws in the experiments are discussed with ways to improve the experiment. The hypothesis is accepted, modified, or rejected in this section. The "take home message" and significant points about the experiment are emphasized.

Literature Cited or References: Throughout the paper the author(s) may cite other papers and experiments. Instead of giving a full citation, the author name and year of publication are given in parentheses. Citations are necessary to document where author(s) information is coming from and to give credit to other researchers. At the end of the paper, the Literature Cited section provides the full reference for the reader. This way the reader can check into similar experiments that were performed or obtain more information. The format for the citations is determined by the editors of the journal. The hyperlink to the journal online may be include.

OBJECTIVE

After this lab, students will know how to better comprehend scientific articles.

NOTES

None

Part 1. Reading Overview.

Read through the scientific journal article provided by the instructor. You may not understand everything, but try your best to understand what is written. Answer the following questions about article.

1. What is the title of the article?

2. In which journal did the article appear?

3. Who are the authors?

4. What year was it published?

5. There are six main parts to the paper (not counting the title). List them each.

6. How many figures appear in the article? How many tables?

7. How were the control cities determined?

8. What cities in Ohio were included in the study?

9. What are the common names of the following? *Agrilus planipennis, Sitta canadensis*

10. What are the scientific names of the following? Red-bellied woodpecker, hairy woodpecker

11. What are the effects of the emerald ash borer on the birds in this study?

Part 2. Parts of the Scientific Paper

Review the six main parts of the paper. Name the part of the paper that matches the descriptions below. You can just list the answer next to each letter (no need for a complete sentence).

1. This section includes the actual data collected and observations made. This section describes what happened during the experiment. ____________________

2. This section is a summary of the entire experiment. A quick overview of the objectives, methods, results, and conclusions can be found here. ____________________

3. This section provides the complete list of references that were used in the paper. Each reference was referred to in a simpler form somewhere in the paper. This section must include author, date, title, page numbers, and volume of the reference. ____________________

4. This section explains how the experiment was performed and how the data was collected. ____________________

5. This section states the objective of the experiment and provides some background information about the study or justifies why the experiment needs to be performed. ____________________

Reference

Campbell and Reese. (2014). *Biology* (9th ed.). San Francisco: Pearson/Benjamin Cummings Publisher.

lab 17

Gel Electrophoresis (Virtual Lab)

INTRODUCTION

Electrophoresis is a technique used to separate different sized molecules (such as protein or DNA). Samples of the molecules are applied to a gel and then a current is run through the gel. The smaller fragments will move farther than the larger fragments. The fragments can be compared to a known sample to draw conclusions.

OBJECTIVE

After this lab, students will know how to explain the basics of gel electrophoresis.

Visit the virtual lab created by the University of Utah.

http://learn.genetics.utah.edu/content/labs/gel/

1. Run through the simulation to introduce yourself to gel electrophoresis.
2. Below the simulation, read through the "Can DNA demand a verdict?" section which explains how electrophoresis is used by investigators.

QUESTIONS

1. Why is a buffer added to the gel?

2. What causes the DNA to move through the gel?

3. Which pieces of DNA move the farthest through the gel? Why?

4. Explain how gel electrophoresis can be used in a criminal investigation.

5. Explain another use for gel electrophoresis.

6. Provide a short critique (one to two paragraphs) of the simulation. Did you learn the basics of electrophoresis? What were the strengths of the simulation? Weaknesses?

References

Campbell and Reese. (2014). *Biology* (9th ed.). San Francisco: Pearson/Benjamin Cummings Publisher.
University of Utah. http://learn.genetics.utah.edu/content/labs/gel/

lab 18

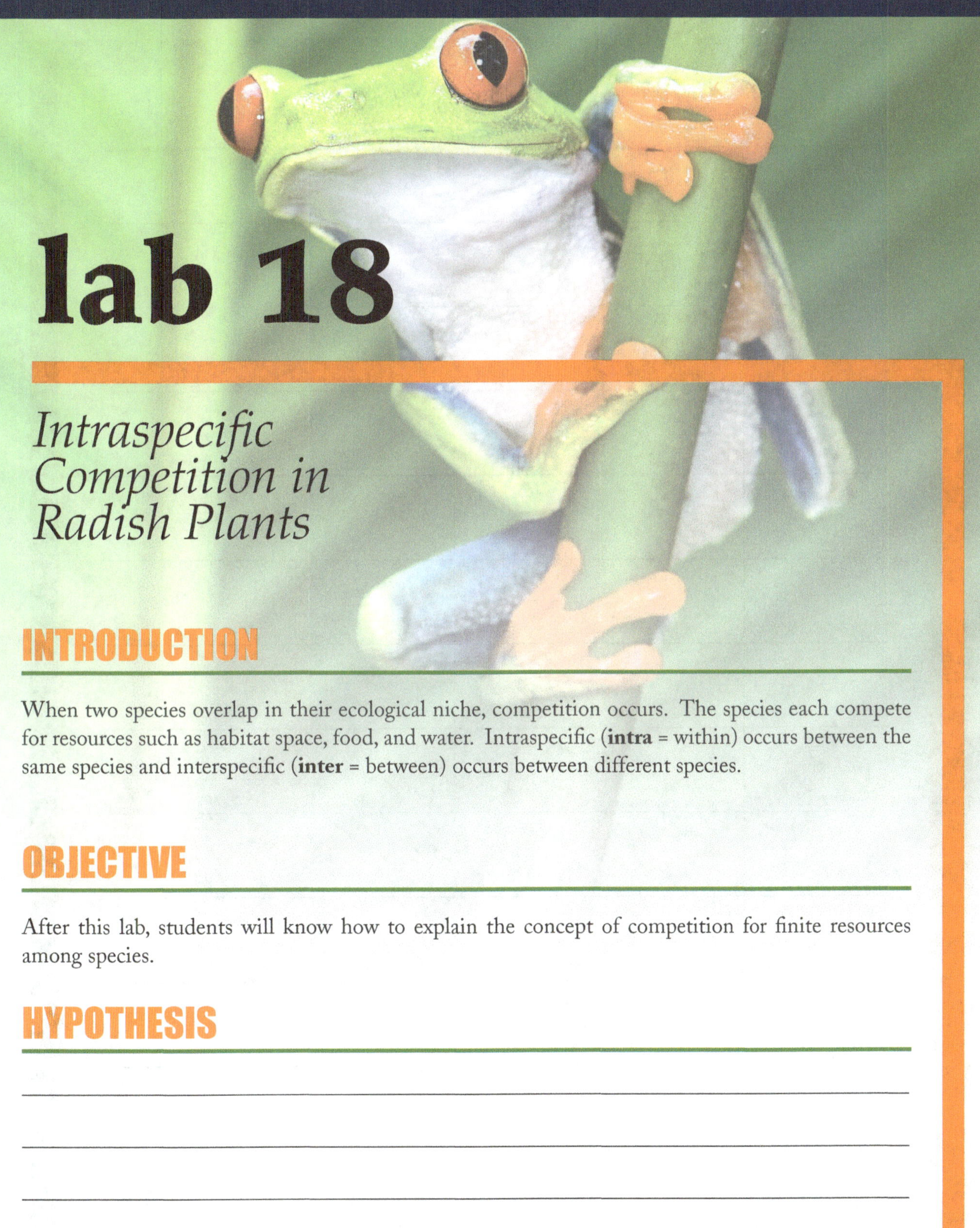

Intraspecific Competition in Radish Plants

INTRODUCTION

When two species overlap in their ecological niche, competition occurs. The species each compete for resources such as habitat space, food, and water. Intraspecific (**intra** = within) occurs between the same species and interspecific (**inter** = between) occurs between different species.

OBJECTIVE

After this lab, students will know how to explain the concept of competition for finite resources among species.

HYPOTHESIS

____

____

____

____

MATERIALS

- Radish seeds planted one month prior to this lab

PROCEDURE

Planting

1. About 1 month prior to data collection, you will be planting radish seeds that vary in the number of plants per pot (n = 3).
2. After 1 month, shoot weight and root weight will be collected for each experimental group.
3. Use 5-inch pots to plant three replicates of the following number of seeds. (Plant a few extra in case germination does not occur. After sprouting the plants can be thinned.)
4. Label the pots R1-1, R1-2, R1-3, etc., to indicate radish, the number of seeds, and the replicate.
5. Plant the following with three replicates. (This equals 12 pots total.)
   a. 1 plant per pot
   b. 2 plants per pot
   c. 4 plants per pot
   d. 8 plants per pot
6. Pots will be maintained in the greenhouse until harvest day.

Data Collection

7. Gently remove the shoots (stems and leaves) from each pot and rinse off the roots from each pot.
8. Towel dry the roots.
9. Weigh the shoots and the roots from each pot separately and obtain the average weight.
10. Then determine the root-to-shoot ratio by dividing the average root weight by the average shoot weight.
11. Record your data in Tables 14 through 16.
12. Create a line graph in MS Excel of the average total weight (roots + shoots) of pots versus the number of radish plants in the pots.
13. Create a line graph of the root-to-shoot ratios of pots versus the number of radishes in the pots.

CALCULATIONS

Table 14 Average root weight of radishes in 5-inch pots

Radishes/pot	Rep. 1 (g)	Rep. 2 (g)	Rep. 3 (g)	Average
1				
2				
4				
8				

Table 15 Average shoot weight of radishes in 5-inch pots

Radishes/pot	Rep. 1 (g)	Rep. 2 (g)	Rep. 3 (g)	Average
1				
2				
4				
8				

Table 16 Root-to-shoot weight of radishes in 5-inch pots

Radishes/pot	Root Weight (g)	Shoot Weight (g)	Total Weight (g)	Root/Shoot Ratio
1				
2				
4				
8				

Graphs:

Name ____ Date _____________________

QUESTIONS

1. From the line graph of total weight, is there evidence of intraspecific competition in radish? Explain.

2. From the line graph of root-to-shoot ratios, is there evidence of intraspecific competition in radishes? Explain.

3. Did the results support or reject your hypothesis? Explain.

4. How did competition affect the development of radishes?

5. For what resources are the radish plants competing?

Reference

Campbell and Reese. (2014). *Biology* (9th ed.). San Francisco: Pearson/Benjamin Cummings Publisher.

lab 19

Ubiquity of Microorganisms

Microorganisms are virtually everywhere around us. They are in the air around us, all over and inside our body, and on the surface of virtually every object around us. By definition microorganisms are too small to be seen without the aid of magnification, but we can grow them in visible amounts.

With this lab we will demonstrate the ubiquity of these microorganisms (particularly prokaryotes) by growing them on agar media (stored in Petri dishes). This lab will also demonstrate the need for aseptic techniques. These techniques, which we will learn throughout the semester, will help us minimize contamination in the future.

Microorganisms form colonies on Petri dishes after incubation. Each colony indicates a microorganism that has landed on the plate and reproduced. Bacteria undergo a rapid asexual reproduction known as binary fission. Therefore, each colony of bacterium is millions or billions of bacteria that are genetically identical to the one you initially caught on the plate. Bacterial colonies come in all types of shapes and colors but are typically small. You might also find fungal colonies on your plate. These colonies tend to be larger and hairy.

NOTES

HYPOTHESIS

____

____

____

____

MATERIALS

- Trypticase Soy Agar (TSA) plates
- Sterile polyester-tipped applicators

PROCEDURE

1. Day 1: Each group of four students will need five TSA plates.
   a. Plate 1: Expose to the air for 30 minutes.
   b. Plate 2: Divide into four sections by drawing a grid on the bottom of the plate. Each person should touch one sector of the plate with a finger.
   c. Plate 3: Get a volunteer to kiss the plate.
   d. Plate 4: Swab some part of your body and streak the plate.
   e. Plate 5: Group's choice. Go find some microorganisms on campus.
2. Incubate all five plates upside down at 37°C in the class incubator.
3. Further notes:
   a. These plates are sterile and the air is filled with bacteria. When you streak a plate, be sure to do it quickly to avoid prolonged exposure to the air. You should always hold the plate upside down until you streak it. Never open the plate and let it sit on the lab bench.
   b. Swabs are also sterile, so the tip of the swab should only come in contact with the area you are sampling. Avoid touching the tip of the swab with your hands and don't let the swab touch anything else. Remember, bacteria are everywhere. It is easy to contaminate your samples.
4. Day 2: Observe the colonies growing on your plates and fill out Table 17.
   a. Estimate the number of colonies on each plate.
   b. Estimate the number of different kinds of colonies.
   c. Describe two colonies from each plate, noting color, texture, margin, elevation.
   d. Refrigerate your plates and save them for Lab 2 (Microscopy) and Lab 3 (Gram Stain).
   e. Draw a plate in Figure 12 and record observations.

Table 17 Colony Characteristics

Environment Sampled	# of Colonies	Diversity	Color	Texture	Margin	Elevation

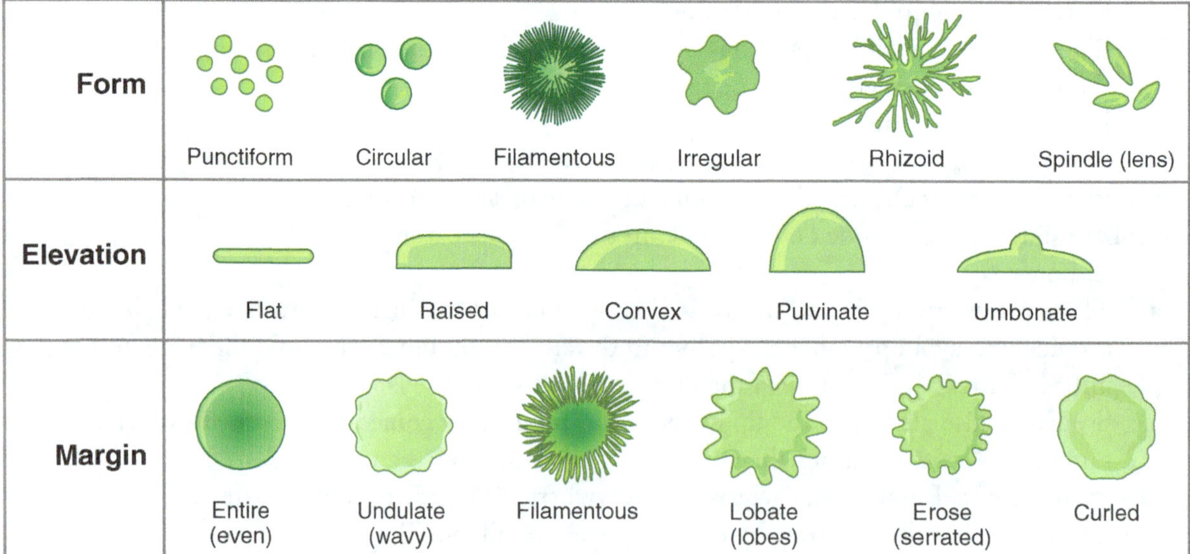

Figure 12 Ubiquity of Microorganisms Plate
© Kendall Hunt Publishing, Co.

CONCLUSIONS

QUESTIONS

1. Why did we incubate the plates at 37°C?

2. Why did we invert the plates in the incubator?

3. Why do we only write on the base of the plate?

4. Name some ways we can reduce bacterial contamination.

References

Reprinted with permission from Clark and Green, Kendall Hunt Publishers, Dubuque, IA.
Bauman, Robert. (2013). *Microbiology with Diseases by Taxonomy* (4th ed.). Glenview, IL: Pearson Education.
http://inst.bact.wisc.edu/inst/index.php?module=book&func=displayarticle&art_id=119

lab 20

Simple Stains and Microscopy

INTRODUCTION

All living organisms can be divided based on their cell structures. Smaller bacterial cells without true nuclei are referred to as prokaryotes. Bacteria come in three basic shapes. Cocci are spherical cells, bacilli are rod shaped, and spirilla are thread shaped. Prokaryotes are typically 0.5 to 1.5 µm in diameter. This small size requires us to use oil immersion (1000x TM) to properly get the cells into focus. We will look at several prokaryotic bacteria.

OBJECTIVES

After this lab, students will know how to:
1. Use theory of and operate a basic light microscope
2. Use the oil immersion objective lens
3. Display techniques of slide preparation and basic staining
4. Explain basic prokaryotic anatomy

- 5 plates from the previous Ubiquity of Microorganisms lab
- Disposable glass slides
- Wire loop
- Bunsen burner
- Methylene blue
- Prepared slide
  - Bacteria: 3 types

PROCEDURE

Focusing the Prepared Slide

1. Place prepared slide of Bacteria 3 types on your microscope. Notice there are three separate samples on the slide. You will have to focus on each of these individually.
2. Observe the slide at 1000x TM (oil immersion).
3. Draw each sample in Figure 13. Indicate the morphology (shape) and grouping of your bacteria in the space provided under the pictures.
4. THOROUGHLY remove all oil from the lens and microscope after seeing all three samples.

Heat-Fixed Smear

5. To prepare your own slide, place a very very small drop of water on the slide (one circle only please).
6. Using a sterile loop (heat the loop until it glows red, allow it to cool for 10 seconds), scrape a barely visible amount of a bacterial colony from one of your Ubiquity plates. Make sure you choose a colony that is separated from the others and try to scrape only one colony. Place the scrape in the drop of water and use the loop to mix the bacteria and water into an emulsion. Your emulsion should appear slightly cloudy (if not, you did not get enough bacteria from the colony). Try not to leave any lumps (If this is impossible just ignore those areas of the slide when you view it through the scope). If your emulsion does not spread out well on the slide and beads up instead, your slide is dirty and oily. Allow the spread emulsion to dry. This should take only 5 minutes unless you used a large drop of water. To speed the drying, pass the slide over the flame to warm it slightly. **BE CAREFUL**, the slide should never get too hot while the bacteria are still in liquid form or the liquid will boil and burst the bacterial cells. Touch the slide to the back of your hand and verify that it is only warm and not hot.
7. **Heat fix** the bacteria by passing the slide over a flame three or four times. The slide should be hot but not too hot. If you cannot touch the slide to the back of your hand without great pain then it is too hot. (*Do not heat-fix until the slide has dried…this will cause the bacteria to rupture and ruin the stain.*)
8. Place the slide smear up on the bench top to cool for a few minutes. Do not proceed to step 7 until the slide has completely cooled. *Failure to wait for the slide to cool will result in the stain crystallizing in the slide.*

9. Stain the slide covering the entire heat-fixed smear circle with methylene blue (about 5–7 drops). Stain for 30 seconds. **Never ever allow the slide to dry with stain on it as this will also result in crystals**. Then rinse the slide by spraying with an indirect stream of water. This is one of the critical parts of staining…too much water and you will wash away your bacteria…too little and the stain is too dark and hard to look at. It just takes practice to figure it all out. Blot the slide dry with a paper towel (not lens paper) Do not rub the slide, only blot it.
10. Observe the slide at 1000x TM (oil immersion).
11. Repeat steps 5–10 with a second colony on a fresh slide.
12. Diagram and label both bacteria in Figures 14 and 15. Indicate the morphology and arrangement of your bacteria in the space provided under the pictures.

Tips for Finding Bacteria with the Microscope

1. You should be able to see your heat-fixed and stained smear on the slide as a light haze with your eye. If not then you should use more bacteria next time.
2. Position the slide on the stage with the smear side up (T in the upper right corner) centered under the lens.
3. Find the sample on 10X. This is relatively easy and there is no way to smash the lens on 10X so move it up and down fairly quickly until you see your sample come into focus.
4. Verify that you are focused on the sample by moving the slide back and forth while looking through the scope. It is easiest to move the slide back and forth with one hand while you focus up and down with the other. If nothing is moving then you are focused on lens dust. Try to refocus on the bacteria. If you cannot find them, try focusing on the wax from your circle, then move into the center of the circle slowly to find the bacteria.
5. Once you are in focus revolve the nose around so that you are between 40x and 100x. **Do not adjust the focus.**
6. Place a drop of oil on top of your stained smear and swing the 100x into position. The 100x lens should make contact with the oil. WATCH TO SEE THAT THE LENS DOES NOT HIT THE SLIDE WHEN YOU SWING IT INTO POSITION. If this happens it will scrape your sample away. You should have no problem as long as your 100x is screwed into the scope completely. Check this before you proceed. Also be sure the 100x is fully extended (sometimes the safety spring sticks).
7. Using the fine focus knob ONLY, move up and down in both directions until you find your sample. It should not be more than two or three turns in either direction. Move the slide back and forth to be sure you are focused on bacteria and not lens dust. Focus on the wax if you cannot find anything and then move into the circle to find the bacteria.
8. Other helpful hints:
   a. Do not worry if the view on the 40x objective does not get perfectly clear. Get it as best you can, then you are ready to move to the 100x oil immersion lens.
   b. Always be aware of the position of your lens relative to the slide and try not to smash them together. The 100x safety spring should never be depressed.
   c. If you're moving the slide and the image in the scope is still, you are focused on lens dust.
   d. The more bacteria on the slide, the easier it is to find them on the scope.

Figure 13 Bacteria 3 types prepared slide 1000x TM

NOTES

 Unknown bacteria stained with methylene blue from Ubiquity of Microorganisms lab 1000x TM

NOTES

Figure 15 Unknown bacteria stained with methylene blue from Ubiquity of Microorganisms lab 1000x TM

NOTES

Name ____ Date _______________________

1. Describe how bacterial cells differ from our cells.

2. Which lenses on the microscope must be used with oil?

3. Why do we need to stain bacterial slides?

4. Should a slide with liquid bacteria on it be heat fixed? Why/why not?

References

Reprinted with permission from Clark and Green, Kendall Hunt Publishers, Dubuque, IA.
Bauman, Robert. (2013). *Microbiology with Diseases by Taxonomy* (4th ed.). Glenview, IL: Pearson Education.

lab 21

Epidemiology

INTRODUCTION

Media headlines often talk about disease outbreaks. These diseases may be local (the chicken at the church dinner had *Salmonella*), national (flu incidences across the country), or worldwide (Ebola virus). The public health system is responsible for coordinated efforts at infectious disease control. For example, each state has a communicable disease section of its public health laboratory that collects and examines culture specimens and data from local physicians, health departments, hospitals, and epidemiologists (health professionals who identify patterns and causes of disease). Findings are shared with other agencies in the state, the Centers for Disease Control and Prevention (CDC), and the World Health Organization (WHO).

Diseases can spread through a population through various means. *Contact transmission* is the physical meeting of a source and a new host, often spread through person-to-person transmission. Touching, shaking hands, and kissing are all common examples of this kind of contact. Other examples include direct contact with secretions or body lesions (such as that with herpes and boils), intimate or sexual contact (sexually transmitted diseases), and transmission from mother to infant via breast milk (staphylococcal infections). Many disease organisms such as *Salmonella* and *Campylobacter* can even be transmitted through *direct contact* with animals, animal eggs, or animal products.

Indirect contact, a second common mode of disease transmission, refers to transmission from a source to a new host through an intermediary—chiefly an inanimate object. Thermometers, eating utensils, drinking cups, and bedding are all common intermediary objects. *Pseudomonas* bacteria are commonly transmitted in this manner.

A third mode of transmission, *droplet spread,* occurs when a pathogen is carried on a particle larger than 5 μm. Because this is a large particle, it quickly settles out of the air; therefore, *droplet spread* depends upon the proximity of a new host to a pathogen source, such as a person sick with measles. Sneezes and coughs typically generate the droplets.

OBJECTIVES

After this lab, students will know how to:
1. Explain the way transmission of disease occurs throughout a population
2. Use data to trace the origin or original carrier of a disease

HYPOTHESIS

____

____

____

____

____

____

____

NOTES

MATERIALS

- 0.001 M HCl
- 1 M NaOH
- 0.04% Phenol Red solution
- 13 mm test tubes
- Tube racks
- Transfer pipets

PREPARATION (to be completed by instructor/lab prep. staff prior to class)

1. Using letters, label the 13 mm test tubes to match the number of students in the class. If you have a small class, you should include "phantom" students in this exercise.
2. Put 1 mL of HCl in all but one tube. These are the uninfected.
3. Put 1 mL of NaOH in the remaining tube. Note the letter of this tube. This represents the infected individual.

PROCEDURE

1. Distribute the lettered 13 mm test tubes, one per student.
2. Each student takes four 13 mm test tubes. Label each tube 0, 1, 2, and 3.
3. Each student puts 5 drops of the solutions into the tube labeled "0". SET ASIDE.
4. Find a partner and "swap fluids." Record the letter and name of your partner on the data sheet. Using your larger, lettered tubes, each person should draw up some of their own fluid and place 5 drops of their own fluid into their partner's tube. Cap and shake your tubes.
5. Before moving on to the next partner, take a sample of your fluids (5 drops) and put this in the 13 mm test tubes labeled "1". SET ASIDE.
6. Repeat steps 4 and 5 two more times, but use a new 13 mm test tube each time (the ones labeled "2" and "3" respectively).
7. Add 1 drop of Phenol Red to each student's large 13 mm test tubes and record the result in the last column of your data sheet. Yellow is uninfected and pink-purple is infected.
8. Add 1 drop of Phenol Red to the 13 mm test tubes if tracking the data becomes difficult.
9. Record data in table 18.
10. From the collected data try to determine which tube was infected.

Table 18 Epidemiology class results

Student Letter	Name	Partner 1	Partner 2	Partner 3	Result (+) or (-)
A					
B					
C					
D					
E					
F					
G					
H					
I					
J					
K					
L					
M					
N					
O					
P					
Q					
R					
S					
T					

CONCLUSIONS

Name ____ Date _____________________

QUESTIONS

1. Describe how you determined who the original carrier was. Was this process easy or difficult? If you had a class size of 100, how would this process change?

2. Relate the classroom simulation to sexually transmitted diseases. Why is knowing a partner's sexual history important?

3. Research the terms "outbreak," "epidemic," and "pandemic." Define each of these terms and provide an example of each that has occurred in our country or world. Cite your sources in your response.

Reference

Campbell and Reese. (2014). *Biology* (9th ed.). San Francisco: Pearson/Benjamin Cummings Publisher.

lab 22

Protista and Algae

INTRODUCTION

This lab will allow you to examine various animal-like protozoan species. In addition, there are some species of plantlike green algae.

OBJECTIVE

After this lab, students will know how to differentiate between common species of protozoans and algae.

PROCEDURE

Find each species under the microscope, draw it at 400x TM, and then either answer the questions or supply comments about the species.

1. Euglena
    a. Draw and label the Euglena in Figure 16.
    b. Make some comments about the appearance and movement of the organism.
    c. What shape are the Euglena? What characteristic(s) make them plantlike? What characteristic(s) make them animal-like? Answer in the space provided.
    d. Label it with its scientific name.
2. Paramecium
    a. Draw and label the Paramecium in Figure 17.
    b. Label the cilia on your drawing. Make some comments about the appearance and movement of the organism.
    c. Label it with its scientific name.
3. Stentor
    a. Draw and label the Stentor in Figure 18.
    b. Write comments that describe how this organism looks different than the others examined.
    c. Label it with its scientific name.
4. Amoeba
    a. Draw and label the Amoeba in Figure 19.
    b. Can you see cytoplasmic streaming? Watch and comment on the movement.
    c. Label it with its scientific name.
5. Diatoms
    a. Draw and label the different shapes of diatoms in Figure 20.
    b. Comment on the appearance compared to other Protista.
    c. Label it with its scientific name.
6. Spirostomum
    a. Draw and label the Spirostomum in Figure 21.
    b. This is a ciliate with a macronucleus that looks like a string of beads.
    c. Comment on its appearance.
    d. Label it with its scientific name.
7. Volvox
    a. Draw and label the Volvox in Figure 22.
    b. Volvox is a colonial green algae. Each hollow ball contains hundreds of cells.
    c. Comment on its appearance in the space provided.
    d. Label it with its scientific name.

Figure 16 Euglena 400x TM

NOTES

Figure 17 Paramecium 400x TM

NOTES

NOTES

Figure 19 *Amoeba 400x TM*

NOTES

Figure 20 Diatoms 400x TM

Figure 21 Spirostomum 400x TM

NOTES

 Volvox 400x TM

NOTES

Name ____ Date ____________________

QUESTION

1. Which species was your favorite? Why?

Reference

Campbell and Reese. (2014). *Biology* (9th ed.). San Francisco: Pearson/Benjamin Cummings Publisher.

lab 23

Overview of Plants

After this lab, students will know how to:
- Differentiate between different classifications of plants
- Differentiate between male and female gymnosperms

MATERIALS

- Various live or mounted plants
- Light microscope
- Prepared moss slide
- Prepared fern slide
- Male and female pinecone slide

PROCEDURE

Part A. Mosses

1. Observe the samples of living moss (or plastic mounts).
   a. Draw and label in Figure 23. Answer the questions in the space provided.
   b. How large are fully grown moss?
   c. Why isn't moss taller?
   d. Where is the sporophyte located?
2. Observe the moss slide containing moss antheridium and moss archegonium.
   a. Draw and label a picture of each in 100x TM in Figure 24. Answer the questions in the space provided.
   b. Your book may help you in recognizing each. Label the sperm in one and the egg in the other.
   c. Describe the differences between the two.
   d. Why are mosses found close to the ground in relatively damp or humid environments?

Part B. Ferns

3. Observe the living fern.
   a. Draw and label in Figure 25. Label the fronds, fiddleheads, stems, and roots. Look for sori and draw/label them.
4. Observe the fern slide.
   a. Draw and label in Figure 26. Label spores (if available), prothalia, young sporophyte.
   b. Also indicate on your drawings the following: gametophyte, sporophyte, haploid cells, diploid cells.

Part C. Gymnosperms

5. Observe the male and female pinecones (slides).
   a. Draw and label in Figure 27.

6. Cut open a pine nut (lengthwise).
    a. Draw and identify the embryo in Figure 27.
    b. Sample some toasted pine nuts (unless you have a nut allergy).
    c. Describe their taste in the space provided.

Part D. Angiosperms

7. Observe the monocot and dicot stem.
    a. Draw and label in Figures 28 and 29, respectively.
8. Obtain a sample flower and "dissect the flower."
    a. Draw and label in Figure 30.
    b. Include the following labels on your drawings: petal, sepal, anther, filament, stigma, style, ovary.

Figure 23 Moss specimen

NOTES

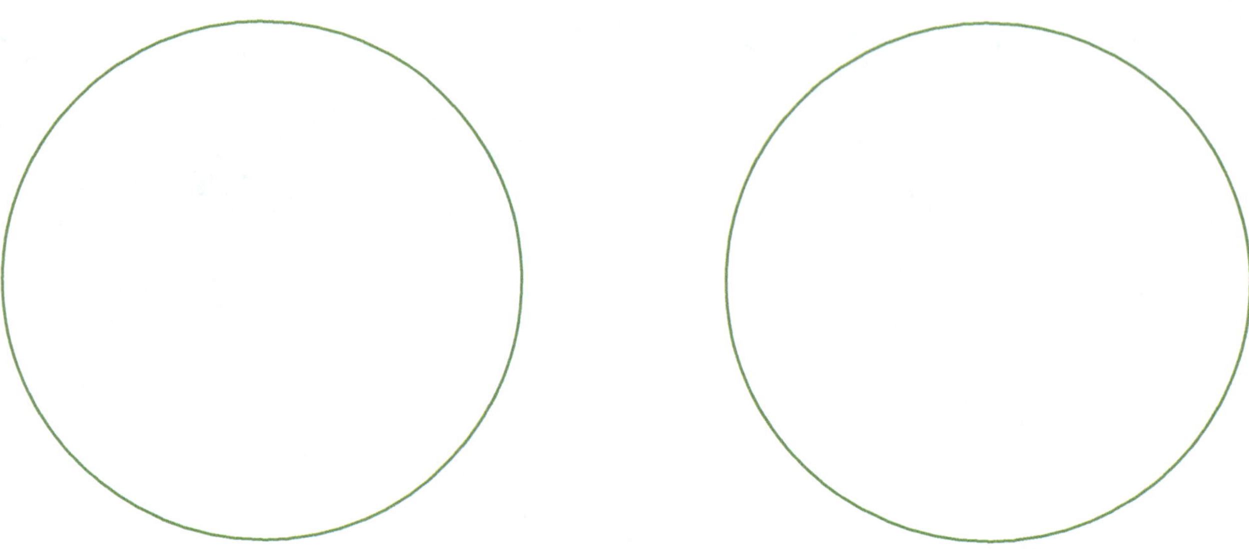

Figure 24 Moss antheridium and archegonium 400x TM

NOTES

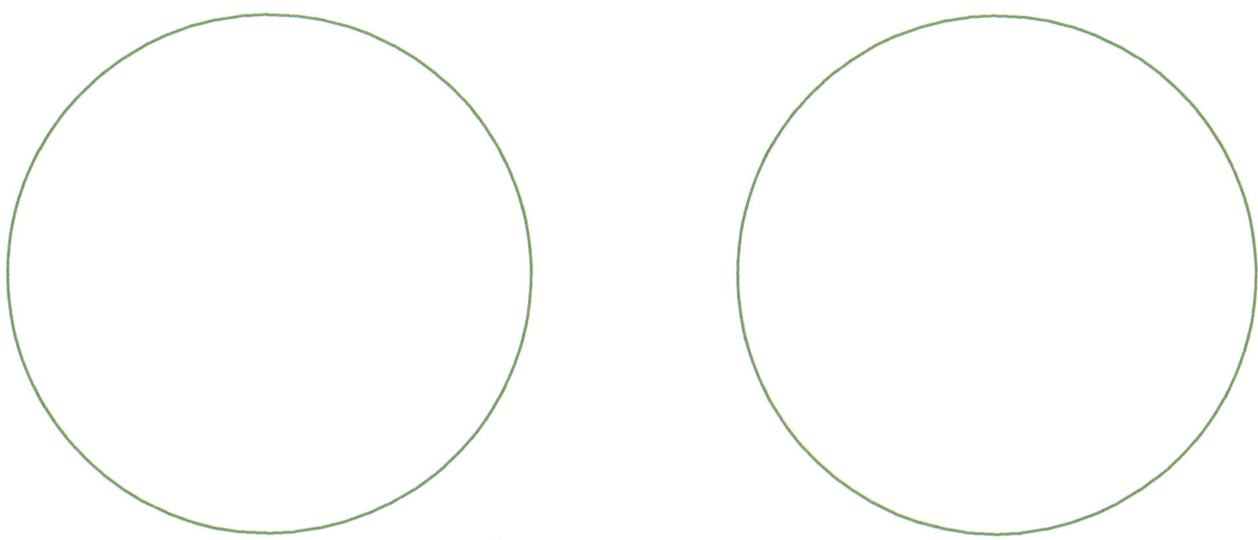

Figure 25 Fern specimen

NOTES

Figure 26 Fern slide 400x TM

NOTES

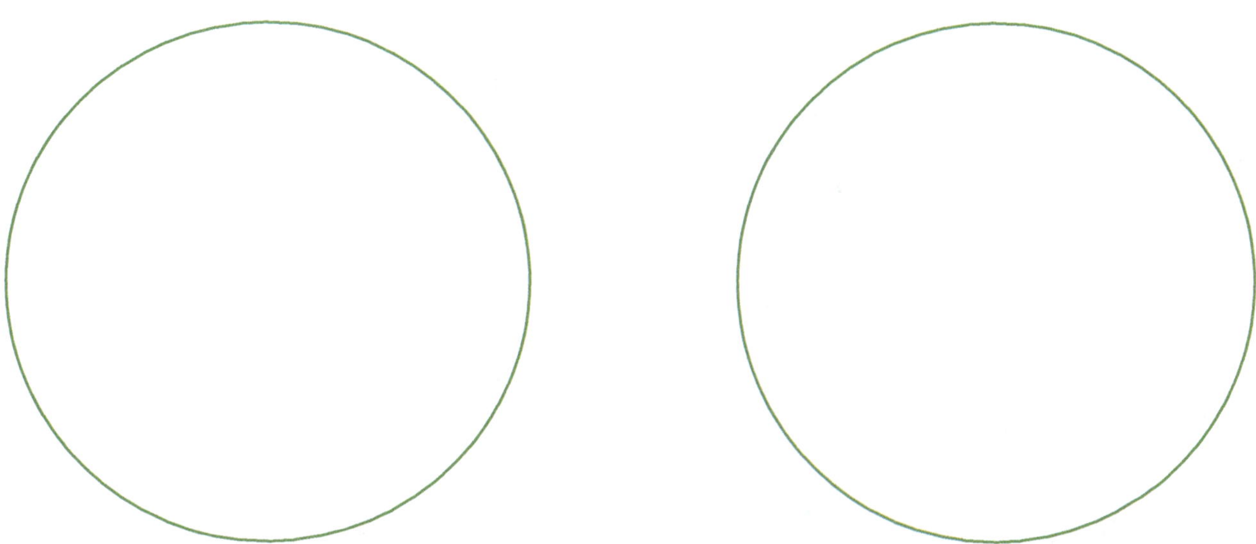

Figure 27 Male and female pinecone 400x TM

NOTES

Figure 28 Monocot stem

NOTES

Figure 29 Dicot stem

NOTES

Figure 30 Flower specimen

Reference

Campbell and Reese. (2014). *Biology* (9th ed.). San Francisco: Pearson/Benjamin Cummings Publisher.

lab 24

Flowers, Seeds, and Fungi

OBJECTIVES

After this lab, students will know how to:
1. Identify the parts of a flower
2. Identify the parts of a seed
3. Identify three different classifications of fungi

MATERIALS

- Fresh flower
- Dissecting scope
- Fungi prepared slide

PROCEDURE

Part A. Flower Anatomy

1. Obtain a sample flower (or if you wish, examine a flower found growing on campus). Use the flower diagram in your textbook to help locate the following parts.
    a. Make a sketch of your flower and label the features in Figure 31. Fill in the functions of each flower part on Table 19.
    b. Give the common and scientific names of your flower.
2. Go to the dissecting scopes and observe the three types of seeds: bean, corn, and pine nut.
    a. Sketch and locate the embryo, seed coat, and food supply for each seed in Figures 32, 33, and 34, respectively. In the case of the bean seed, the food supply is not as visible as the cotyledons, which are the seed leaves. Beans are a dicot (two cotyledons). Dicots generally have seeds that split in half easily.   Label the cotyledons on the bean seed you observe.
3. Think of another example of a dicot: ____

Part B. Fungi

1. Examine the prepared slide showing three fungal types. You may see both sexual and asexual structures. Use your book and the information in class to identify the three types observed. Draw and label each in Figure 35, then try and identify the species observed.
    a. Aspergillus (an Ascomycota)
    b. Penicillium
    c. Rhizopus (Zygomycota)
2. Observe the mushrooms available. Draw and label in Figure 36.
    a. Include Pileus (capped shape with gills on the underside); Ring; Stipe (stalk)
    b. Observe a section of the mushroom under the dissecting scope, and label in Figure 37.
3. You or your instructor will be propping the mushrooms above a piece of filter paper to make a mushroom print. After 1 week, observe the spores that have been released from the gills. Draw and label in Figure 38.

Table 19 Flower anatomical parts

Flower Part	Number Observed	Function or Role
Sepals		
Petals		
Stigma		
Style		
Ovary		
Anthers		
Filaments		

Reference

Campbell and Reese. (2014). *Biology* (9th ed.). San Francisco: Pearson/Benjamin Cummings Publisher.

NOTES

Figure 32 Bean seed observed under dissecting scope

NOTES

GENERAL BIOLOGY LABORATORY MANUAL

Figure 33 Corn kernel observed under dissecting scope

NOTES

Figure 34 Pine nut observed under dissecting scope

NOTES

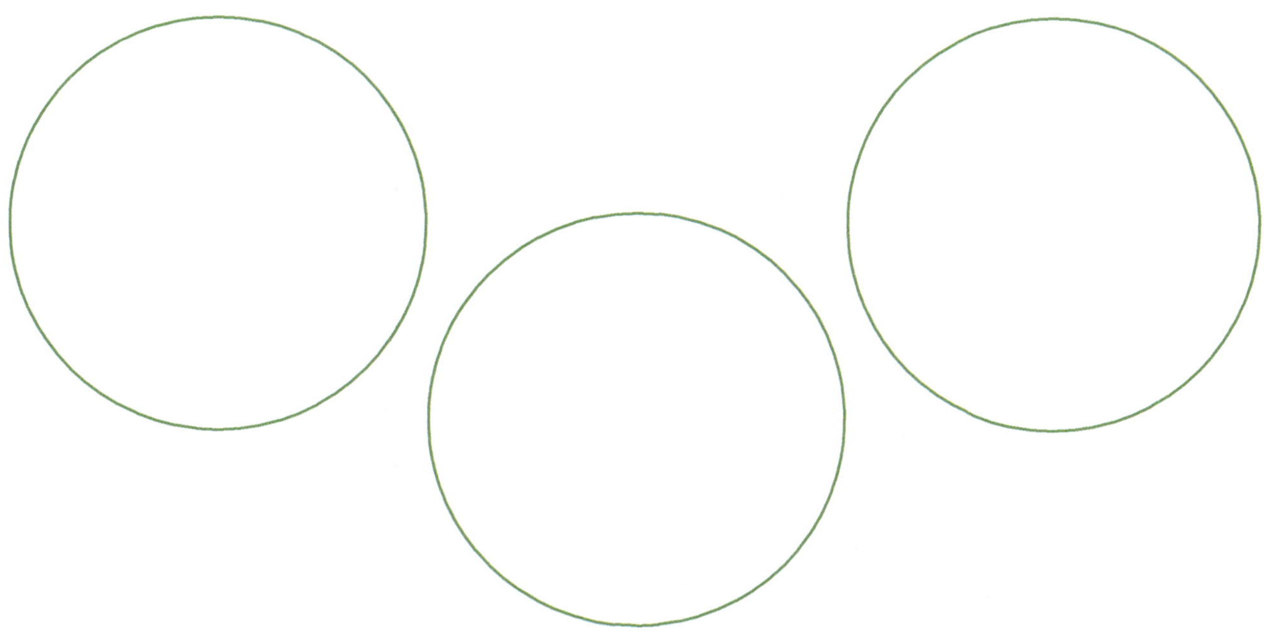

Figure 35 Fungi 3 types

NOTES

Figure 36 Mushroom specimen.

NOTES

Figure 37 Mushroom observed under dissecting scope

NOTES

____

____

____

____

____

Figure 38

NOTES

lab 25

Earthworm Dissection

INTRODUCTION

The earthworm dissection will give you the opportunity to examine a representative of the Animal Kingdom. The earthworm, *Lumbricus terrestris*, is a common organism found in the United States and throughout the world. It belongs to a group of worms called the segmented worms or Annelids. These worms all have true coeloms.

OBJECTIVE

After this lab, students will know how to identify the external and internal anatomical structures of the earthworm.

NOTES

MATERIALS

- Earthworm specimen
- Dissecting kit
- Dissecting tray

PROCEDURE

External Anatomy

1. Draw your worm and label the following features in Figure 39.
   a. Prostomium: very tip of the worm above the mouth
   b. Clitellum: swollen portion important in reproduction
   c. Pores: may be visible and excrete mucus
   d. Setae: bristles on every segment (except the first and last) used for movement (you may notice these by touch rather than vision)

Internal Anatomy

2. Practice cutting at the posterior (tail-end) of the worm. You want to cut lengthwise along the dorsal (back) surface. Your cut should be just through the body wall and not so deep that you cut through the organs. Once you are successful, cut the first 20 or so anterior segments (head) and pin back the body wall. Draw and label the following organs in Figure 40:
   a. The pharynx is generally found in segments 1–6; soil enters the body from the mouth here.
   b. The cerebral ganglion consists of two whitish "knobs" which are the "brain" of the worm.
   c. The esophagus is a narrow channel often covered by the "hearts" and seminal vesicles.
   d. The crop attaches to esophagus at about segment 14 and is connected to the gizzard that is very muscular. Food is ground here.
   e. The intestine is the remainder of the digestive tract.
   f. Seminal vesicles are large, whitish bodies found in segments 9–13.
   g. Seminal receptacles are smaller, rounded projections in the same region.
   h. Worms are hermaphroditic, but the ovaries are difficult to locate without the aid of a microscope.
   i. The dorsal blood vessel is the reddish tube that runs along the mid-dorsal line of the intestine.
   j. At the esophagus are five aortic loops or "hearts."
   k. Moving aside the intestine and looking for a white threadlike structure that runs the length of the worm exposes the ventral nerve cord. One end attaches to the "brain."
   l. Nephridia are looplike structures in each segment used for excretion.

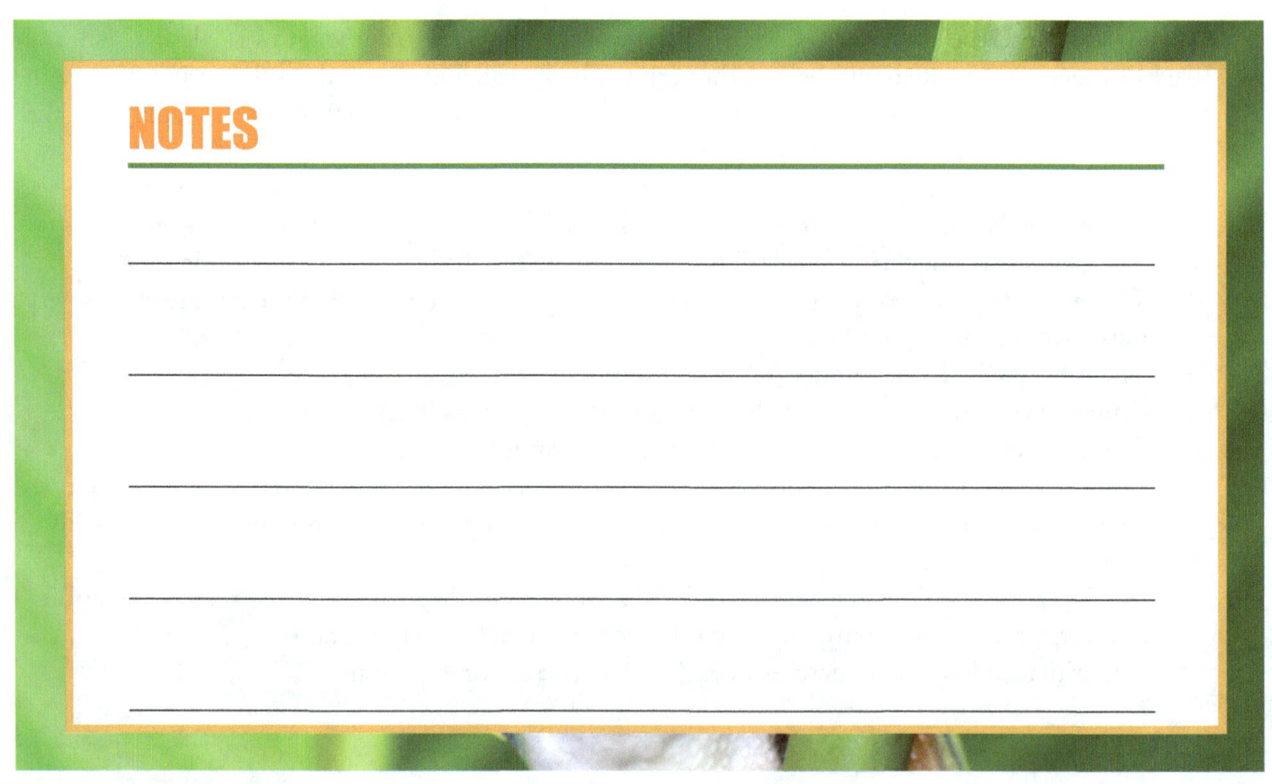

Figure 39 Earthworm external anatomy

NOTES

Figure 40 Earthworm internal anatomy

NOTES

Name ____ Date ____________________________

1. How many segments does your worm have?

2. Does the worm have radial symmetry or bilateral symmetry?

3. Count the number of segments from the prostomium to the clitellum.

Reference

Campbell and Reese. (2014). *Biology* (9th ed.). San Francisco: Pearson/Benjamin Cummings Publisher.

lab 26

Frog Dissection

INTRODUCTION

In this lab, the frog will be used to visualize some of the organs discussed in class. A frog is not a human, but it is a vertebrate; therefore, several anatomical features of the frog are similar to the human.

OBJECTIVE

After this lab, students will know how to identify the internal and external structures of the frog.

NOTES

MATERIALS

- Frog specimen
- Dissecting kit
- Dissecting tray

PROCEDURE

A. External Anatomy

The frog has two basic regions—the head and the trunk. The mouth, external naris (pathway for air), and tympanic membrane (for hearing) are all easily located. The cloacal opening (posterior or tail end) leads to the cloaca. The cloaca is a chamber that receives products from the digestive, excretory, and reproductive systems. This is one way that frogs differ from humans. Open the mouth slightly and feel for little teeth. You may want to cut the jaw to further expose the tongue, esophagus, vocal sac opening, and Eustachian tube openings.

B. Internal Anatomy

1. Digestive and Respiratory Systems
    Place the frog ventral side (stomach) up. Make one cut from the cloacal aperture to open the body cavity. Do not cut so deep that the organs are affected. Next, make two perpendicular cuts so the flaps of skin can be pinned back and the organs exposed. Draw the internal anatomy and label the following italicized parts:

    The *esophagus* takes food from the *mouth* to the *stomach* (enlarged portion to the right). The stomach leads into the *small intestine*, which then leads into the *large intestine*. Notice how the diameters of the small and large intestines are different. The large intestine leads to the *cloaca*. The *liver* is the largest organ in the frog and is generally a dark brownish/red. The liver has a variety of functions including the production of bile. Bile is stored by the *gallbladder* (a small, greenish sac embedded in the middle lobe of the liver). The *pancreas* (whitish organ) sits near the stomach and the small intestine.

    How many lobes does the frog liver have?_________

    Make an incision in the stomach. What does it look like?____

    Air enters the external nares of the frog into the mouth and to the glottis. It passes over the larynx (lots of cartilage), which contains the vocal cords. The bronchi connect the larynx to the lungs (a left and a right).

2. Additional Organs

   The *heart* is resting above the *lungs*. A frog's heart only has three chambers (two atria and one ventricle). The *spleen* is a dark colored, rounded organ located in the coils of the intestine. The *urinary bladder* sits above the cloaca and you may be able to trace the ducts back to the *kidneys* (dark, elongate organs embedded in the wall of the body cavity). *Adrenal glands* are thin bands of whitish tissue on the ventral surface of the kidneys. If your frog is a male, the *testes* will sit above the kidney and look ovoid and cream colored. *Fat bodies* are yellowish and are at the anterior end of the testes. These provide important hormones and lipids for reproduction. If your frog is female, the *ovaries* will occupy much of the body cavity if the frog was captured in spring (filled with eggs). If not, the ovaries will be small and ventral to the anterior end of the kidney. Coiled oviducts carry the eggs to the outside of the frog.

   Is your frog male or female?___________

3. Once you have identified the major organs in the frog, you may want to carefully remove the small and large intestines. Uncoil them and then measure them.

   small intestine length:________   diameter:_________

   large intestine length:________   diameter:_________

4. Mini-quiz: Learn the organs discussed above. Your instructor will point to several organs and have you identify them.

5. As a student, would it have been more beneficial to use a different specimen to study the organs, such as a fetal pig? Why or why not?

Reference

Campbell and Reese. (2014). *Biology* (9th ed.). San Francisco: Pearson/Benjamin Cummings Publisher.

lab 27

Fetal Pig Dissection

INTRODUCTION

Fetal pigs are a good example of mammal anatomy. Although there are some differences, the organs, body plan, and functional relationships between the organs are similar to humans. This lab will give you the chance to integrate all the systems of the body we have studied this semester.

OBJECTIVE

After this lab, students will know how to identify and label the anatomical structures of a fetal pig.

NOTES

MATERIALS

- Fetal pig specimen
- Dissecting kit
- Dissecting tray

PROCEDURE

Follow the procedures given to each group when dissecting the fetal pig.

Write answers to questions below.
Initial when you have completed the steps (use the instructions sheet as a guide).

1. I understand the terms *anterior*, *posterior*, *ventral*, and *dorsal*. ________________________

2. I have identified the following features: ____________________________
   pinna, nostrils, umbilical cord, forelimbs, hindlimbs, thoracic region, abdominal region, anus, tail

3. What is the sex of your pig? ____________________________

4. I have identified the following features from the respiratory system: ____________________________
   trachea, lungs, bronchi, diaphragm

5. I have identified the following features from the digestive system: ____________________________
   esophagus, stomach, liver, gallbladder, pancreas, small intestine, large intestine

6. I have identified the following features from the immune system: ____________________________
   spleen, thymus

7. I have identified the following features from the excretory system: ____________________________
   kidney, ureter, bladder, urethra

8. I have identified the following features from the endocrine system: ____________________________
   thyroid gland, thymus gland, pancreas, adrenal glands, testes or ovaries

9. If my pig is a female I have identified the oviducts and uterus: ____________________________

Be familiar with the organs/glands listed above and the system of the body to which they are affiliated.
You will have a mini-quiz over these features.

Reference

Campbell and Reese. (2014). *Biology* (9th ed.). San Francisco: Pearson/Benjamin Cummings Publisher.

Hiking Notes

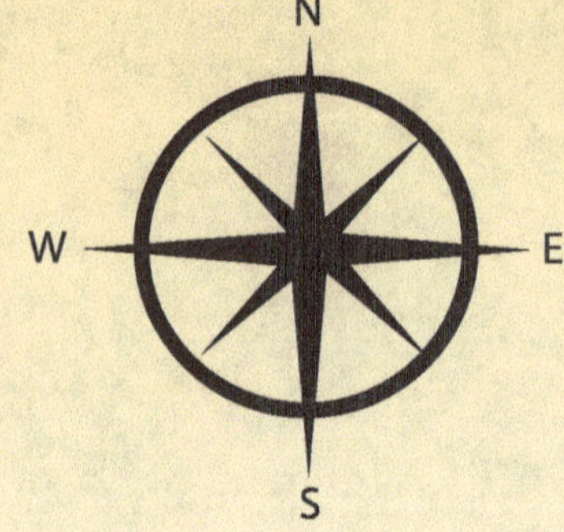

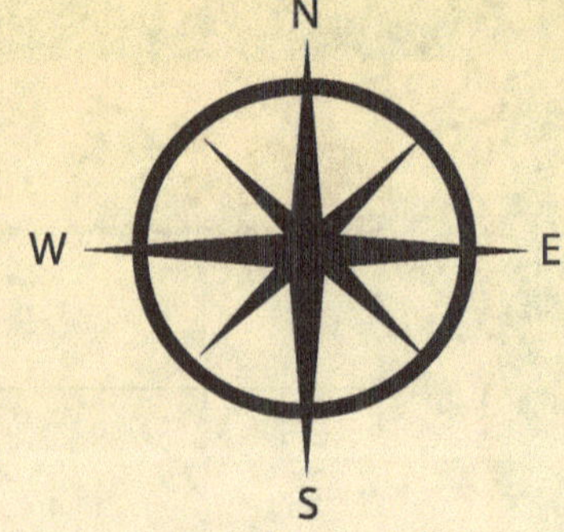

NOTES

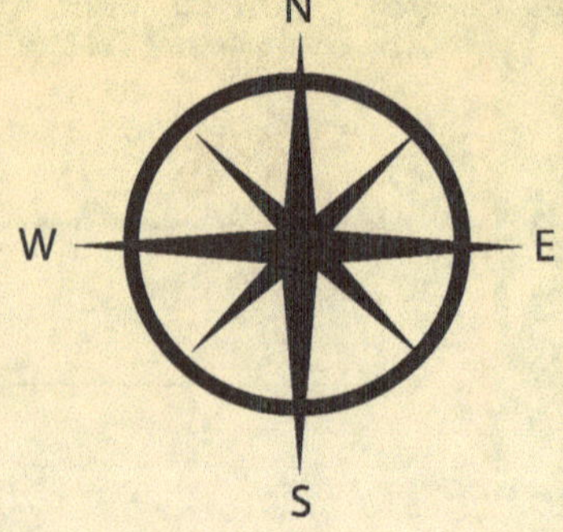

NOTES

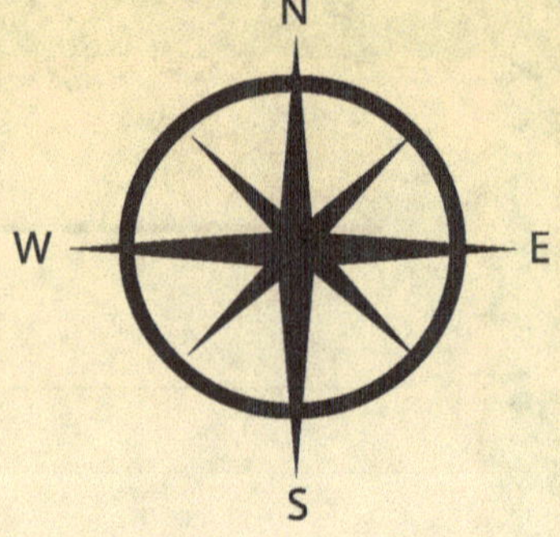

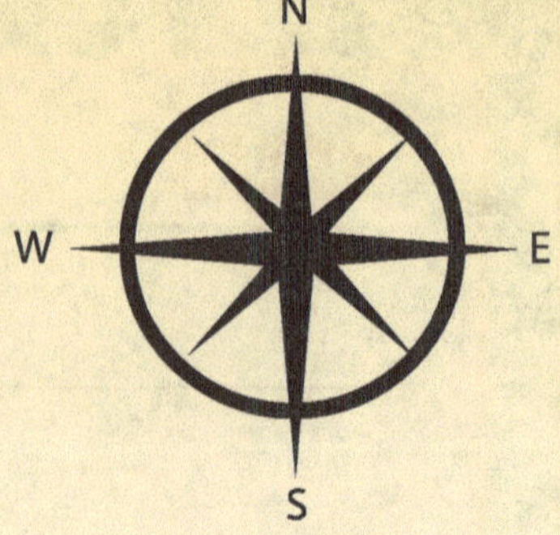

NOTES

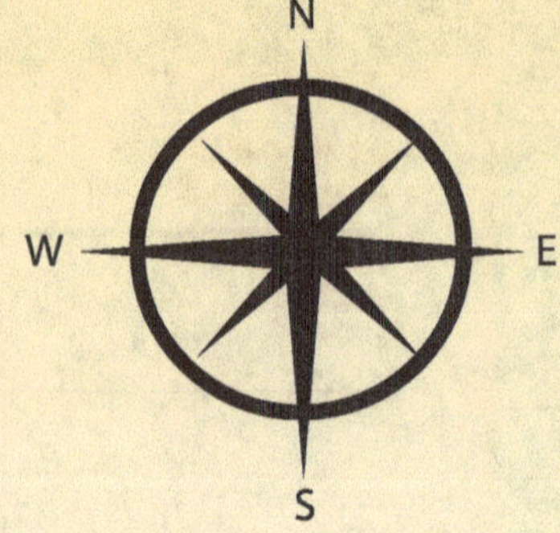

NOTES

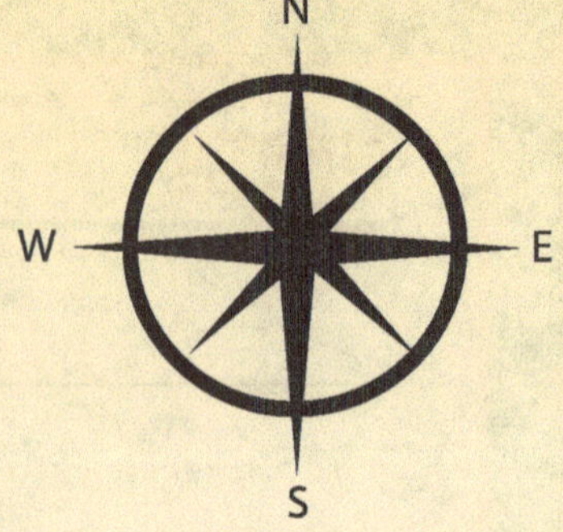

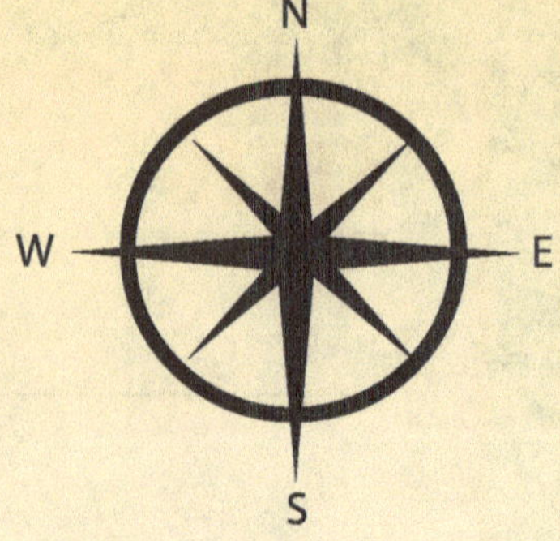

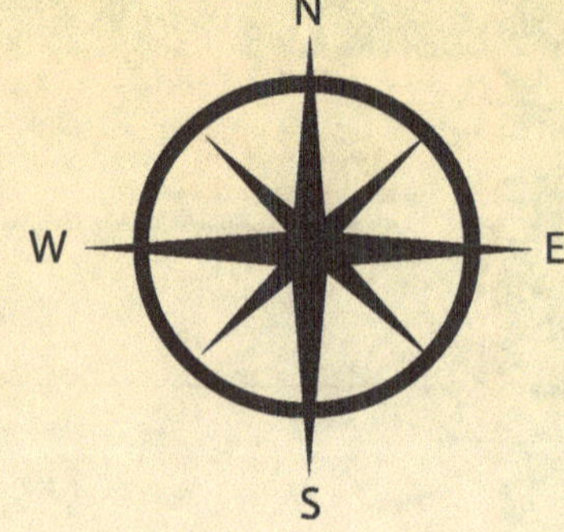

NOTES

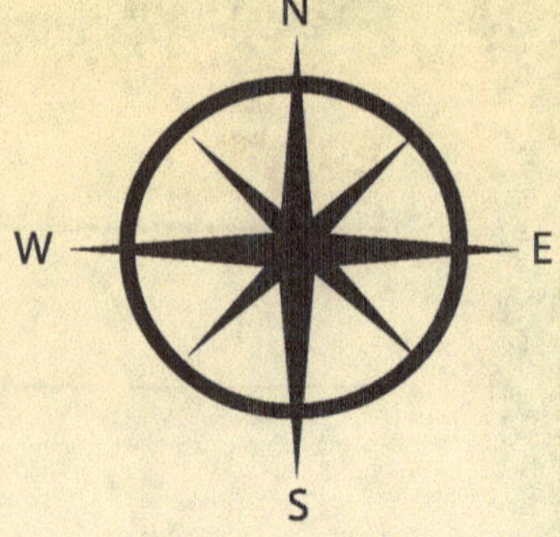

NOTES

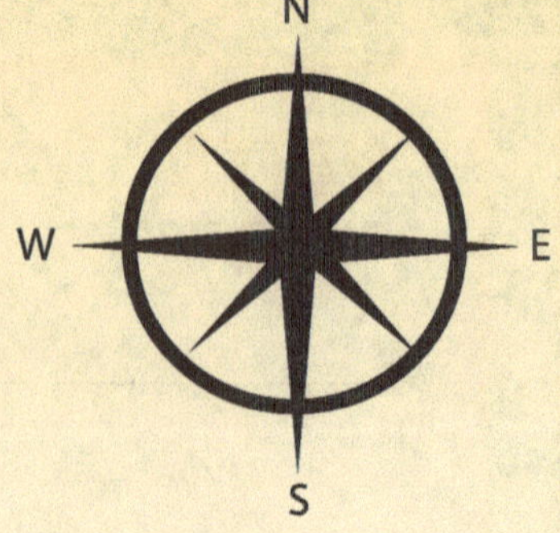

NOTES

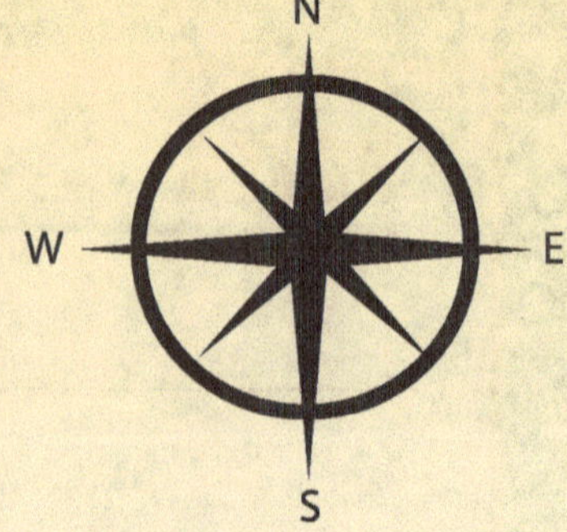

NOTES

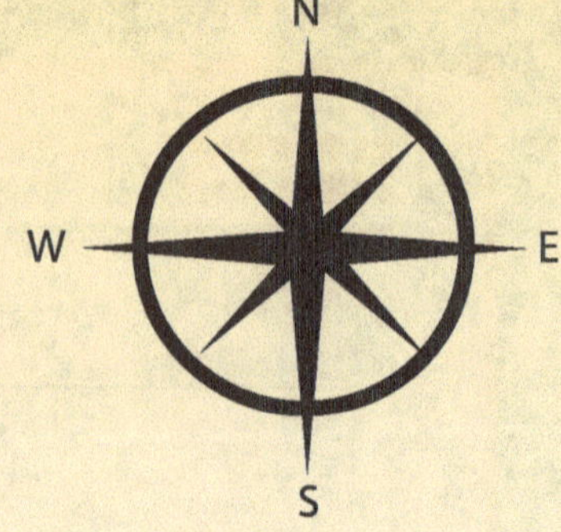

NOTES

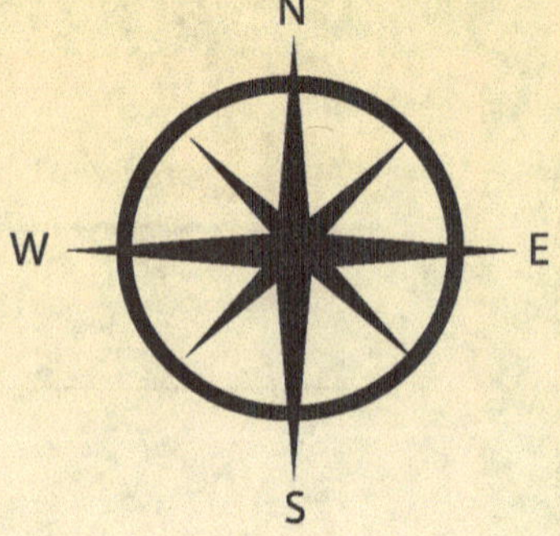

NOTES

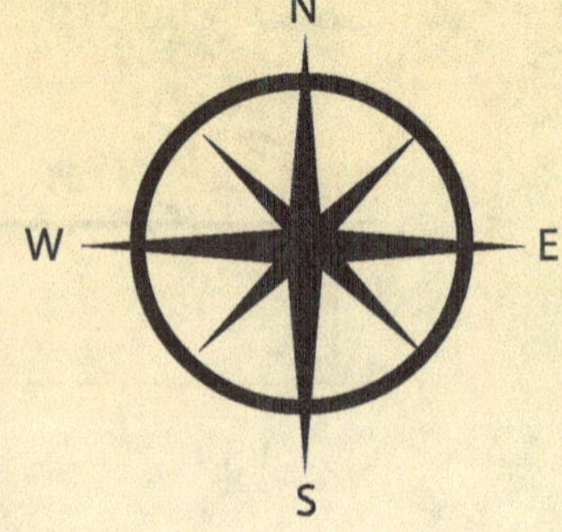

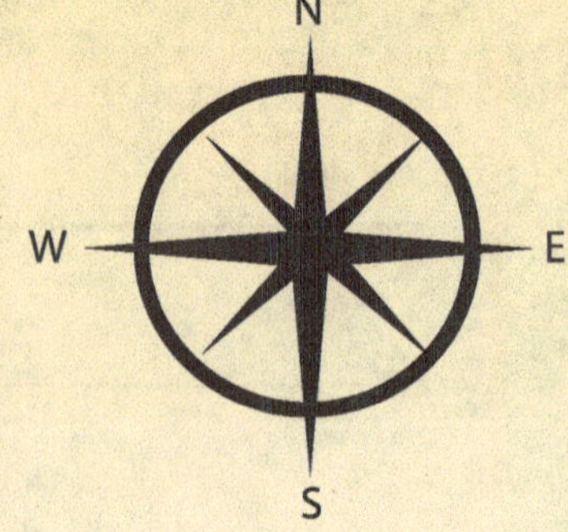